PHYSICAL SCIENCE LABS KIT

Ready-to-Use Activities and Worksheets for Grades 5-9

THOMAS KARDOS

illustrated by
Thomas Kardos, Andi Tanner and Eileen Ciavarella

**THE CENTER FOR APPLIED
RESEARCH IN EDUCATION**
West Nyack, New York 10995

©1991 by

THE CENTER FOR APPLIED
RESEARCH IN EDUCATION

West Nyack, NY

10 9 8 7 6 5 4 3 2 1

Library of Congress Cataloging-in-Publication Data

Kardos, Thomas, 1933-
    Physical science labs kit : ready-to-use activities and worksheets
for grades 5-9 / Thomas Kardos.
        p.   cm.
    ISBN 0-87628-631-7
    1. Science—Study and teaching (Elementary)   2. Physical sciences-
-Laboratory manuals.   3. Education, Elementary—Activity programs.
I. Title.
LB1585.K33   1991
372.3'5—dc20   90-28350
                                                                   CIP

ISBN   0-87628-631-7

THE CENTER FOR APPLIED
RESEARCH IN EDUCATION
BUSINESS & PROFESSIONAL DIVISION
A division of Simon & Schuster
West Nyack, New York 10995

Printed in the United States of America

DEDICATION

This book is dedicated to my wife Pearl, a credentialed teacher, who encouraged, supported and helped me with this challenging task. She was my reader, editor and motivator. My daughter Marla took my portrait for this book, while my children Jennifer and Ron rooted me on for many years in my new authoring venture.

Special thanks to my editor Sandy Hutchison, at Prentice Hall, who took my original book and helped me change it into a ready-to-use, professional resource.
Special thanks to Zsuzsa Neff who was invaluable in the final production.

ABOUT THE AUTHOR

Thomas Kardos, a resident of California, has a B.A. in Mathematics and Physics, an M.A. in Elementary Education and 26 years experience as a classroom physical science and mathematics teacher. He has served as science coordinator and consultant. He co-authored *Physical Science*, a laboratory manual, for Scott Foresman and *FUN WITH PHYSICS AND CHEMISTRY*, which he self-published.

He was the top 1984 finalist, in California, for middle school Chemistry in the national Search for Excellence in Science Education. In 1984 he was an Honors Teacher Award recipient from NASA, NSF, NSTA and the University of Toledo, Ohio. He was also a nominee in 1989 and 1990 for the Presidential Awards for Excellence in Science and Mathematics Teaching.

ABOUT THIS RESOURCE

As a science educator, I faced a common problem. I was searching for a laboratory manual that would provide exciting classroom experiences and solid laboratory literacy for my students as well as ready-to-go, tested activities and reproducible worksheets for me. I wanted a resource that would

- allow teachers to use their own teaching styles;
- emphasize a hands-on, discovery approach;
- provide teacher instructions, forms, and tables;
- require only inexpensive, easy-to-find materials;
- be compatible with most textbooks and simple to use without extensive in-service training or heavy preparation;
- employ an incremental learning process, with thinking skill transitions from the concrete to the abstract.
- be usable at several levels of difficulty.

Ultimately, I wrote my own. *Physical Science Labs Kit* **places in your hands a complete, sequential set of activities for teaching about** *measurement, matter, energy, force and motion,* and *science and technology.* The activities are simple; they work the first time and describe single concepts.

Written at a seventh grade reading level, the reproducible worksheets can be used through the middle and secondary grades (with fine-tuning of the math and vocabulary as needed). These activities also adapt well for ESL and special education students.

You can use the activities as a complete program or pick and choose to supplement your existing curriculum. Each chapter begins with teacher notes, followed by the activity sheets. Materials called for are all relatively inexpensive; most are available in the typical kitchen.

In addition, you'll find useful forms and teaching tips in "Suggestions for the Teacher," including a reproducible blank lab worksheet and a laboratory procedures handout. You'll also find specific information to help you teach the activities in this book effectively.

Rather than focusing on content, this book meets your students' needs to manipulate and explore, to discover for themselves through experimentation, and to invent new uses for their discoveries. With it, you can guide your students toward an enthusiastic mastery of the scientific process.

Happy Sciencing!

Thomas Kardos

SUGGESTIONS FOR THE TEACHER

This introductory section gives you general information and reproducible pages that will help you manage your laboratory time. In addition to important safety information, there are recommendations for classroom behavior, guidelines and techniques that help students build thinking skills. Reproducible pages include a blank lab sheet useful with all the investigations in this book and student guidelines for behavior in the science lab.

SAFETY INFORMATION

1. Follow all safety rules of your school district, your county, your state and all special federal requirements.
2. Provide students with safety goggles, aprons, and gloves when needed.
3. Provide safety instruction.
4. Provide principal with a written statement that general safety instruction has been taught.
5. Specific safety instructions have to be provided at the beginning of each period, before hazards exist.
6. Do not allow students to taste any substance, unless you specifically direct them to do it.
7. Students may use sharp or cutting instruments only if you provide instruction and are assured that students are fully competent in their use. Use of these tools must be done under your closest supervision.
8. Use Bunsen burners, portable or stationary. Instruct students in their safe handling.
9. Do not use alcohol burners.
10. Do not use candles in classroom, unless allowed by your state.
11. Do not allow students to have matches. Use a piezoelectric flame starter, a flat-file gas starter, or safety matches.
12. Do activities which produce toxic gases under a fume hood or outside.
13. Do not allow students near any dangerous or potentially harmful demonstration.
14. Do not allow students to carry any dangerous chemicals or equipment.
15. Warn pupils on the hazards of using and mixing chemicals outside of school.
16. Report promptly to school any laboratory accident, no matter how minor.
17. Keep acid-neutralizing materials close at hand.
18. Dispense chemicals with dropper bottles when possible. (Dropper Chemistry)

19. Do not handle radioactive materials.

20. Keep all hazardous materials under lock and key. Do not permit student access.

21. Check with experienced district personnel on the correct way to handle and dispose of hazardous materials.

22. Be alert to proper ventilation in the classroom.

23. Organize activities to provide smooth traffic flow.

24. Have a written contingency plan for emergencies. Test it.

25. Keep all flammables away from any open flame.

26. Label all containers clearly.

27. Follow local fire ordinances.

28. Keep students at a safe distance from your demonstrations.

29. Use extreme care in handling wet-cell batteries.

30. Use extension cords with ground plugs only.

31. Pull extension cord from wall by the plug, not the wire.

32. Do not handle any electric device which has just been used. It is hot and may burn you and your students.

33. Do not open any electric appliance unless it is disconnected from the power supply.

34. Cover all sharp edges of mirrors, glass, etc., with tape.

DISCLAIMER

The safety rules are provided only as a guide. They are neither complete nor totally inclusive. The author and the publisher do not assume any responsibility for actions or consequences in following instructions provided in this laboratory book.

LABORATORY PROCEDURES

It is important for your own safety and security that you perform laboratory experiments within certain guidelines. Some of the substances used are harmful. You must handle them with care. Most equipment is delicate in order to provide you with precise answers. You need to use it with care.

1. Handle only equipment which is required for your experiment.
2. *Never eat or taste anything*, unless you are told to do so by the teacher.
3. When handling harmful chemicals, you will need to wear a plastic apron and goggles. Goggles must be over your eyes at certain times during the activity. Your teacher will indicate when.
4. After your investigation, clean all equipment. Replace all equipment *exactly* the way you found it.
5. Use and replace the plastic apron with care.
6. Use and replace the goggles with extra care.
7. Should you come in contact with any substances during an investigation, wash the affected areas with plenty of water. Promptly report what happened to the teacher.
8. *Always* fill out the Lab Worksheet as completely as possible. Turn in your lab work before deadline.
9. (additional instructions) _____

_____

_____

_____

_____

_____

_____

_____

_____

_____

_____

_____

Student Signature   Date   Parent Signature   Date

HANDLING MATERIALS

Lay out on a counter or table near a sink all the supplies you need for your current investigations. Assistants can do this for you. Have students go one way only in picking up their equipment. This will speed up preparation time. If necessary, have each partner go separately and get only a part of the total equipment. For returning materials, reverse the traffic pattern.

You will need a sink or several large trash cans for liquid and other waste. Have students go to the disposal site first, before returning equipment. It takes less time if one or two students clean up all equipment. Always emphasize care in handling and returning all science equipment, even an ordinary glass jar.

Maintain several bottles of diluted cleaner in spray bottles for students to use to keep their desks clean. A student is required to have a spotless desk, regardless of who was there before him. If he does not clean it, then he must take responsibility for the mess.

BEHAVIOR

I do not have many behavior problems, for I have simple standards. All students are provided with a copy of my standards, which they and their parents must sign. A clearly printed copy is also posted in the classroom. The entire school has nearly identical standards, which helps. I *demand* that students be involved with science.

In cases of unacceptable behavior, I will provide a warning, have the student write a note home, give time out with a call home, and, finally, suspend the student. A few calls home in the first week of school give you extra needed support. Misbehavior is not tolerable in any formal or informal laboratory setting. Should a student misbehave during lab activities, I will assign him an alternative quiet writing task, with a failure grade for the lab activity.

REWARDS

When students complete their activity, I may allow:

1. Free time to play with science equipment.
2. Free time to study.
3. The use of computers with science software.
4. Help with class equipment.
5. The use of an extensive resource library on every conceivable aspect of science (in back of room: mostly unadopted textbooks).
6. Wandering around to the many centers of interest.

I give out certificates for citizenship and achievement. Several fast-food chains make these available, along with complimentary food. I send positive letters home to parents, and sometimes I call. Positive feedback and praise is the key to good student behavior and performance.

_____ Instructor

CLASSROOM BEHAVIOR EXPECTATIONS

1. Be on time.
2. Bring your notebook, paper, textbook, pencil, and all materials you will need.
3. Do not disrupt others. A disruption occurs when you get the attention of others.
4. Treat all students in class with courtesy.
5. No gum chewing, eating, or wearing of inappropriate garments: hats, gloves, etc.
6. Restroom use is limited to break time between class periods. Exceptions are made for emergencies or medical reasons.

CONSEQUENCES!

ONE INFRACTION: You are warned. This is only a warning.

TWO INFRACTIONS: You must write a note home, advising your parents of your improper behavior. The letter must be signed by your parent(s) and returned the next day. If the letter is not returned, you will be referred to the administration, who will contact your parent(s) immediately for a conference.

THREE INFRACTIONS: The teacher will contact parents on the same day. You may get time out in another class.

FOUR INFRACTIONS: You will be suspended from class and you may return only after your parents have conferred with the teacher and a school administrator.

NOTE: NO BEHAVIOR MAY BE DISCUSSED DURING CLASS TIME.

_____   _____
STUDENT SIGNATURE   DATE   PARENT SIGNATURE

TEACHING THE LAB ACTIVITIES

1. Review and teach math skills students need.

2. Demonstrate the activity by executing it for the class. This modeling is necessary to allow for questions and discussions.

3. Early on, provide the text for the *purpose, procedure,* and *equipment* sections of the lab worksheet (which follows).

4. Good instructions are crucial to student success.

5. Hand out equipment and have students practice the activity until they understand the procedures.

5. After going over any necessary safety precautions, supervise the students as they conduct the activity. Some students will not understand the activity from just reading the procedure. Help them do it and then have them read it and explain the procedure to you.

6. At the beginning, assist students in filling out the lab worksheet by dictating the text. Keep it simple. Help with the setup of the data table (most models for the data table are supplied in their activity sheets). Gradually have students take over the writing of lab reports.

LAB WORKSHEET

\# _____

_____

Name   Period

EXPERIMENT TITLE:

_____

Date

_____

PURPOSE: The purpose of this investigation is to _____

_____

_____

PROCEDURE: _____

_____

_____

_____

_____

_____

_____

_____

EQUIPMENT:

QUANTITY	NAME	QUANTITY	NAME
_____	_____	_____	_____
_____	_____	_____	_____
_____	_____	_____	_____
_____	_____	_____	_____
_____	_____	_____	_____

DIAGRAM: (Label all parts.)

DATA:

CONCLUSIONS: _____

_____

_____

_____

_____

_____

_____

_____

_____

BUILDING THINKING SKILLS

Students are building thinking skills as they write conclusions for the lab report. The following breakdown of steps may be helpful to you:

1. Ask that students report their findings in the data boxes as the first step in the writing of conclusions. This step should be possible for all levels of students.
2. Ask that students report their findings in the data boxes and break down the data as *greater than* ($>$), *less than* ($<$), or *equal to* ($=$) some reference data value.
3. Ask that students review the *purpose* or *hypothesis* of the investigation and see if the outcome in the data box supports it. If it does not, how could one restate the hypothesis and redo the investigation?

EVALUATION OF LAB WORK

Initially make sure that students pay close attention to the correct format. Do not be overly concerned with content. After the format is mastered, place emphasis on content. Continue, however, to insist on quality presentation from students: clean papers, neatly written, correctly spelled, in a labeled folder.

When grading for content, always focus on *Data, Graphs,* and *conclusions,* all of which are higher-order thinking skills. *Purpose, Procedure,* and *Equipment* become routine and do not require higher-order thinking skills. After a while you will be able to glance at *Conclusions* and feel the pulse of your class. If you discover that students have trouble with a specific activity, reteach it. If it still is a lost cause, drop it and move on, then try it again a few days later.

I normally give lab worksheets from the first few activities an *A,* if students provide a written report on the data in *Conclusions.* Subsequently the lab grade drops to a *B* unless they add, in *Conclusions,* at least three comparisons of numerical values in data. After a few additional activities, the *A* drops to a *B* unless they can:

1. Report on data.
2. Provide three numerical comparisons.
3. Relate outcomes to hypothesis and/or purposes.

You will need to model several conclusions with the class. Steps 2 and 3 above involve higher-order thinking skills and may not be obtainable from all students. This permits a *C,* as the minimum grade, for the majority of students. For ESL or special education students, remove the math and use the activities only. Ask for descriptions, concrete tallies (like marks on the blackboard), discussion, and little or no written work. You can measure progress from these students by their involvement in the activity.

Grade as quickly as possible. If something is not right, have students do it over at once, for an immediate higher grade. I grade on the spot, as soon as students finish their labs. This allows for personal help, leads to better work, and pleases the student.

Give lots of positive feedback. It makes a big difference if all students begin your course with success. Be generous and give *As* for *Attributes* activity. Success breed success—even low ability and special education students blossom.

TABLE OF CONTENTS

DESCRIPTION

TEACHER'S SECTION

In this chapter students will learn how to describe objects and events more thoroughly after a quick review of the scientific method. A vocabulary standard is introduced, with a list of *attributes*. The concepts of *reference object* and *relative position and motion* are developed. *Measurement,* as a means of accurate description, is introduced later.

1.1 THE SCIENTIFIC METHOD

This activity is for review.

The *scientific method* has definite steps. A problem of some type starts the action. There may be many solutions to the problem, but one sets a limit and makes an educated guess, a *hypothesis*. The hypothesis is tested by *experiments,* whose *purpose* is to collect *data* (observations and measurements.) Data becomes the *evidence* for a *conclusion.* At this point one sees if the data support the hypothesis; if not one revises the hypothesis and starts the entire cycle again. In science it is typical to come up with more questions than one starts with.

This activity supplies you with a chart of the *scientific method.* Make a large poster for your class. Refer to it every time you do an activity. Point out to students the parts which they are executing. I spend most of my time on *data* and *conclusions.*

1.2 ATTRIBUTES AND VOCABULARY

Ask students to bring you spare buttons from home. Actually, you may use almost any objects, as long as all students have the same class of object to describe. Now is the time to discuss with students that many words used for science have new and different meanings. Provide new and different scientific meanings for some familiar words. Next, assign a format for vocabulary presentation. It may be advisable at the start for you to provide students with simple definitions as models. When you do so, use simple words which tie in with concepts. Understanding is more important than formality.

Demonstrate as many words as possible in class. Take a piece of paper, declare it a *system,* then tear it into pieces. Ask if you still have the same system. Define a *system* as a set or collection of parts. It remains the same as long as you do not add or subtract anything from it.

Take one small piece of paper away, showing that now you have a different system. Next, stretch a rubber band. Your system has not changed, since you have neither added or taken away anything. Provide students with the thought that *collection* in English, *set* in math, and *system* in science have close, almost identical meanings.

Vocabulary: Attribute, hypothesis, purpose, investigation, data, evidence, conclusions, system

1.3 ATTRIBUTES LIST

Have your students keep this list at the beginning or end of the science section of their notebooks. From now on, anytime they need to describe something they must refer to this list and use as many attributes as possible. Broad terms, such as *large, medium,* and *small,* may be acceptable at first. You will later cover measurements as tools of science.

1.4 POSITIONS OF OBJECTS

Start by asking the students to describe the position in the room of an object or person, relative to a reference object such as himself, the clock, the blackboard, a globe, a light, or a desk.

Practice this activity until most students have learned to describe the relative positions of objects. The sequence in which students report is not important. The three coordinates plus distance are crucial to the learning experience. Introduce the idea that the descriptions of positions in space are called *coordinates*.

Vocabulary: Reference, relative, coordinate

1.5 OBJECTS IN SPACE AND MOTION

You may wish to use a small stuffed animal or doll in place of the cardboard. The object is a projective DESCRIBER. I call it Mr. DE. Regardless of space configuration, Mr. DE's position and orientation can be found, because:

1. he has a larger right arm,
2. his front is striped, and
3. he has a face.

Class Activity: Place a small piece of string or ribbon in front of the box or book where Mr. DE is, and move this box forward. Place another Mr. DE to the side, leaning against an upright book. Now you have two Mr. DEs looking at each other and reporting. First the moving Mr. DE reports before, during, and after passing the stationary Mr. DE. Next the stationary Mr. DE reports on the positions of the moving Mr. DE. This will duplicate conditions of motion, as when two cars are moving past each other.

When Mr. DE is moving, he sees objects moving toward him, for he uses himself as the reference object. A person in a car sees the road moving toward him. A person on the side of the road may report the car moving toward him, then away.

1.6 UPS AND DOWNS

Class Activity: Have students describe the *position* of many objects in the room, through the eyes of Mr. DE. When students reach a level of good descriptive understanding, change the activity. You think of an object and provide only one descriptive term. Now students must provide the remaining descriptions while guessing which object it is. Turn the reporting Mr. DE upside down at times. Everything Mr. DE sees when upside-down is upside-down from what students really see. They may have trouble with it. Point out that his *up* is their *down,* for example.

When selecting reference objects in space, use the window of an airplane or the console of the Space Shuttle. Spacecraft and space probes use reference stars.

Notice that nearly all people generally define up and down with reference to themselves. Eye level acts as the point of separation between up and down.

Vocabulary: Pole (earth's), separation, planet

1.1 THE SCIENTIFIC METHOD

Purpose: The purpose of this lesson is to review the scientific process.

Information: The scientific process is the operational model which all scientists use.
You start with a *problem.* You follow it up with information, observation, and study. The best possible educated guess you arrived at, is called the *hypothesis.*

The next step is to set up an *experiment* to test the hypothesis. Experiments have steps to follow, a *procedure.* During the experiment, observations are made and collected as *data.* Data are usually organized in a formal manner, such as graphs or tables.

Data are interpreted and checked to see whether they support the *hypothesis.* If they do, the experiment ends. Otherwise one goes back, modifies the hypothesis, and the testing process resumes. Often, in science, researchers come up with more questions than they had originally.

Example: You need to know if a marble rolls down faster on a ramp with a smaller or a larger incline. You try to remember what happens when you play with marbles. You recall playing with ramps. Your *hypothesis:* the steeper ramp will give a marble more speed. You set up an *experiment.* You roll marbles down a ramp supported by two books first, then four. You measure how far the marble rolls. You discover that every time you use four books, the marble rolls three feet farther. Your *conclusion* is that the steeper ramp rolls the marble further and therefore faster. The problem *ends.*

THE SCIENTIFIC METHOD

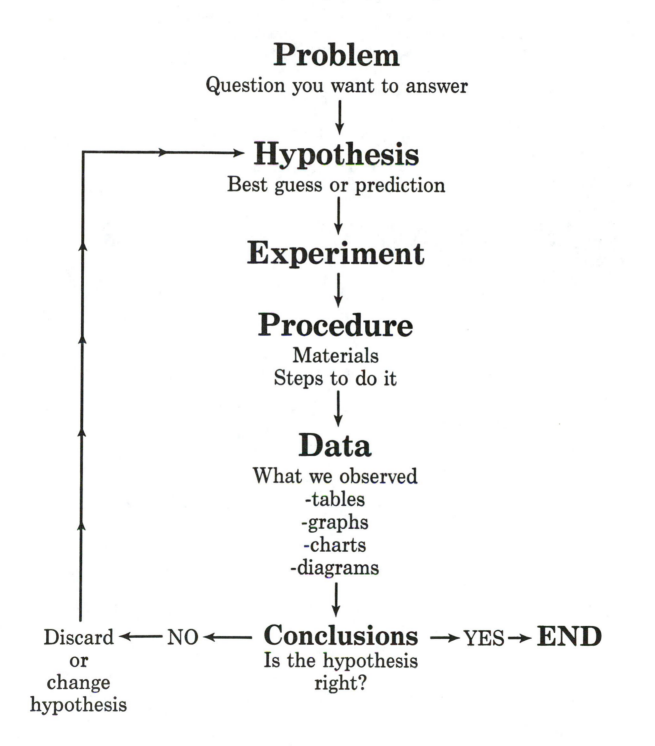

Observation
Made at every step of your experiment

Problem
Question you want to answer

↓

Hypothesis
Best guess or prediction

↓

Experiment

↓

Procedure
Materials
Steps to do it

↓

Data
What we observed
-tables
-graphs
-charts
-diagrams

↓

Discard ← NO ← **Conclusions** → YES → **END**
or   Is the hypothesis
change   right?
hypothesis

1.2 ATTRIBUTES AND VOCABULARY

Purpose: The purpose of this activity is to become more competent in describing objects and situations. You will use attributes (descriptive terms) while practicing the scientific method.

Information: Much time in science classes is used to describe materials, objects, or events that happen during activities, investigations, and experiments. To describe, you use words like *red small light round and dry* etc. These descriptive words are called *attributes*.

Equipment: Button or other teacher supplied object.

Procedure: Take a sheet of paper and number the first eight to ten lines. Add more numbers if you need them. Next, describe a button's attributes, writing one attribute per line. See how many different attributes you can come up with to describe the button. Tape this button to your sheet, since you will need it for another investigation.

Information: Your page, with your attributes, is a *data sheet*. It contains real information you get and record during a science investigation. In science we say that data are the evidence for your conclusions or findings. This is quite different from opinions not based on real facts. Investigations or experiments help us find facts which will provide evidence for our conclusions. Conclusions will either support our most intelligent guess (hypothesis) or suggest that we are incorrect and need to start again.

Vocabulary: Starting today, list all new science words which you come across. List them in your notebook in a vocabulary section. Here is a format that will help you understand new words:

SAMPLE VOCABULARY ENTRY
HOT—Having a high temperature.
   I enjoy hot meals, because heat brings out the many delicate flavors in foods.

FORMAT
1. Offset the writing to the right, so the vocabulary word stands out.
2. Provide a sentence to show a meaning for the new word.
3. Use a *because*-clause within the sentence.
4. Add to this list any other words which are new to you.
5. Underline the vocabulary word.
6. Review new words as soon as you come to class, so you will understand and know them.

1.3 ATTRIBUTES LIST

Purpose: To develop a working set of attributes. These can be changed to suit specific needs.

Information: *Attributes are words or terms which describe properties of material objects.* The attributes listed below are just a beginning. Add more as needed. The more attributes you have, the better you will do! Develop the habit of *always* reading this list while describing objects. Be careful to choose only those attributes which apply in your specific situation.

Note: Never taste or eat anything unless the teacher tells you that you may.

Procedure:

1. Find the DATA sheet from EXPERIMENT #1.
2. Draw a line below your first attributes.
3. Number the page again from one to twenty. You can use the backside.
4. Take the same button used in activity # 1.2 and repeat the activity using the ATTRIBUTES LIST from activity 1.3 as a CHECKLIST.
5. Read each attribute, see how it applies for your object and comment on the DATA sheet.
6. Compare your improvement both in number of attributes and quality of description.

Attributes:

FORMAL OR INFORMAL DESCRIPTION—brief sentence

STATE OF MATERIAL (solid, liquid, gas)

CLASS OF OBJECT (Examples: appliance, machine, fastener, item of clothing, chemical)

PICTURE (Draw a picture; it is worth a thousand words. Show top, bottom, side)

MEASUREMENTS (length, width, height, or other measurements)

MATERIAL (Examples: wood, paper, plastic, metal, food, rock)

TEMPERATURE (hot, cold, or precise measured temperature)

MASS (How heavy? Measure it or compare it to something else.)

NUMBER (How many?)

BULK/VOLUME (How much space does it occupy? How many gallons, quarts, pints, liters?

COLOR (Examples: red, violet, chartreuse)

SEE-THROUGH (*transparent* or *opaque*)

MAGNETIC (Attracted by magnet?)

CONDUCTS ELECTRICITY

BUOYANCY (Does it float? On top, bottom, or middle of liquid?)

SOUND

TASTE (Examples: sweet, salty, acid, bitter, sour)

EDIBLE (food, nonfood)

ODOR (Describe or compare smell.)

HARDNESS (soft, hard, spongy)

FLEXIBILITY (a little, hardly, not at all)

MOISTURE (dry, moist, slimy)

TEXTURE (Examples: smooth, rough, bumpy, cracked)

HOLES (Does it have holes? Where are they: top, bottom, middle, on side, on edge? How many holes? How far apart are they? Describe them.)

HOLLOW (Where?)

EDGE SHAPE (Examples: rounded, sharp, flat)

LIKENESS TO OTHER OBJECTS (Specify.)

FRAGILITY (Can you break it? Where, how, why?)

BEHAVIOR (Does it roll without being touched? Spin? Bounce?)

FLUORESCENCE (Does it glow under black light?)

USES (How is it used under everyday conditions?)

OTHER USES (Invent new uses for this object.)

CLEAN-DIRTY (Describe.)

pH

DOES IT HAVE LAYERS? (Describe them.)

SHINY/DULL

1.4 POSITIONS OF OBJECTS

Purpose: The purpose of this investigation is to become competent in describing objects and their positions relative to a reference object.

Information: If you wish to describe the position of an object in a classroom, you need to define its position *relative* to a *reference object*. You can use general terms or, better yet, perform actual measurements.

When describing the position of an object, you need to specify four items relative to your reference object:

1. above or below it
2. to the right or left of it
3. in front or behind it
4. distance from it
   a. touching
   b. very near
   c. near
   d. far
   e. very far.

1.5 OBJECTS IN SPACE AND MOTION

Purpose: The purpose of this activity is to learn how to describe positions of objects with reference to a known reference object or person.

Information: Up to now, you have described all objects with reference to yourself or another object. Now you will need to become the voice for another observer. You will be provided with a small figure, cut from a piece of cardboard. You will report what this little figure sees.

Equipment: Two small figures, cut from cardboard with clearly marked faces, a larger right arm (coloring it helps). Stripe the front of the figures and tape them to the top of a small box or block.

Procedure: Mr. DE uses himself as the reference object.

1. Place Mr. DE on your table, facing you. He describes where the teacher's table is.
2. Place Mr. DE on the floor facing your chair. He describes where the room door is.
3. You walk toward the blackboard, with Mr. DE in your hand, facing the board. You stop, and he describes what he saw while you were moving. He also comments on you.
4. You hold Mr. DE upside-down facing away from you. He describes the position of your shoes.
5. Lean one Mr. DE (*B*) against the edge of a standing book. The other Mr. DE (*A*) is being pulled along by a string attached to his box or base. Both Mr. DEs are facing each other.
   I. *A* is before *B*.
   II. *A* is directly opposite *B*.
   III. *A* is past *B*.

Give Mr. DE's at B report for positions I, II and III. See the diagram below.

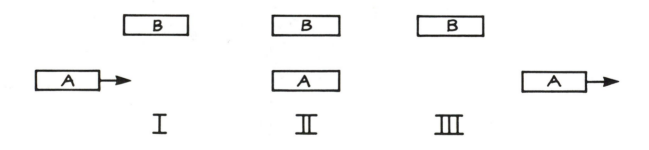

1.6 UPS AND DOWNS

Purpose: The purpose of this investigation is to continue improving personal description skills by taking a closer look at the attributes *up* and *down*.

Information: Whenever someone tells you to look up, you look in the direction above your head. When someone asks you where down is, you automatically point to your feet. This definition is accurate when you use yourself as the reference object. What would happen if you flew out into space and looked up: Are you looking at Earth or into deep space? What if you were flying in a plane and saw a bird and wanted to tell someone else the bird's position?

Equipment: A globe, two Mr. DEs, a piece of string, a plumb bob or any object which can hang as a weight.

Procedure:

1. Place one Mr. DE on the North Pole of the globe, the other Mr. DE on the South Pole.
2. Mr. DE (North Pole) reports where down and up are.
3. The other Mr. DE, on the South Pole, also reports where down and up are.

Questions:

1. What is happening? Which way is really up? What is the reference object?

_____

_____

_____

2. Now pretend that you are in a spaceship and you want to point to where up and down are. How will you do it? How will you describe positions if you, like the Space Shuttle, fly in earth orbit upside down, with your cargo bay facing the Earth?

_____

_____

_____

_____

_____

_____

Information: On Earth people define down as the direction where a plumb bob would point. This direction is toward the center of the earth. It is important to know where down is when one drills for oil—wells are never drilled up. When contractors build structures, they need to define *up,* *down,* and *level.* Contractors use a *level,* an instrument with a bubble in a liquid (usually oil), with parallel sides and set inside a wood or metal box.

Question:

3. You and your friend stand right next to each other. You both fly straight up to the moon overhead. Would you land on the same place on the moon? Explain.

_____

_____

_____

_____

_____

2

MEASUREMENTS

TEACHER'S SECTION

The activities in this section cover:

1. The centimeter
2. The meter and kilogram
3. Using a thermometer
4. Celsius = Fahrenheit conversions
5. Graphing
6. The liter and graduated cylinders
7. Volume: math and measurement
8. Averages and percentages
9. Solids displace liquids
10. The metric system
11. Conversion between metric units

2.1 THE CENTIMETER

Class Activity: Have all students point with their fingertips (not a pencil) to 25 cm on their rulers, then 28 cm, then 17 cm, and many more points chosen at random.

You should select numbers greater than 12, since the 30 cm rulers usually have 12 inches marked on the opposite side. By starting with numbers greater than 12, you automatically help students identify the correct side of the ruler to use for metric measurements. When everyone comfortably identifies and recognizes the centimeter scale, provide values like 15.6 cm, 18.7 cm, etc. Have everyone check with their neighbors for accuracy. Point out that the middle point, .5, is always marked larger, to help in reading the scale. When doing their first measuring activity, students will have different objects and different reference points. Answers will vary. While students measure the same objects, the difference in measurements points to the need for a standard of measurement.

Vocabulary: Centimeter, decimal, metric

2.2 GETTING YOUR HANDS ON A CENTIMETER

For real learning, your students need a working (concrete) experience with some metric units before learning how the entire system operates. Should you disagree, go directly to activity 2.18.

Caution: Students often use the left edge of ruler as the reference point in measurements, rather than the zero mark. In many cases the real zero does not have a 0 symbol printed. Point this out to your students. Measure the lengths of each line and record in your book.

Vocabulary: Millimeter

2.3 I AM A METRIC CREATURE

You will need to place five to ten strips of masking tape on tables, closets, and floor. Label them with a marker *A, B,* etc. Make several strips longer than 1 m so that students will need to cooperate in their measurements. Specify one particular door to measure. Now is a good time to begin the discussion on *precision* in measurements. Have students report their measurements of any one object. You will be amazed at their disparity. Bring up the need for all students to measure from the same reference points. Talk to them on the value of repeating a measurement when accuracy is a need. On this occasion, open the discussion of mathematical averages. This is a simple way to get a middle value and a powerful step in reducing errors in measurements. Average, as an example, a dozen student measurements of tape strip *A*: add the twelve measurements and divide the sum by 12, and you will have the average. Now go and measure it yourself! You will notice that this average value will be reasonable and close.

Vocabulary: Measurement

2.4 MY HEIGHT AND MASS

Nail to the wall, a closet, or elsewhere in the classroom two meter sticks, one above the other. Call this a height measuring station. It will make your work much easier if you have three to five such stations. The key to a reasonably accurate measurement is to have the measuring meter stick horizontal on top of the student's head or perpendicular to the ruler on the wall. Demonstrate this a few times. Students tend to read directly and forget that there is a lower meter stick on the wall. Usually their readings must be added to 100 cm (provided you want answers in centimeters). Metric countries use the centimeter for height.

Obtain a metric bathroom scale. Cover pounds with masking tape. *Variables* are introduced here, so students will notice them. Their height and mass change in relation to the time of the day, as time goes on, etc. Sometimes several variables will come up, but it is crucial that you focus on only one. Most students at this level cannot successfully handle two variables. You are starting a long-term project. Begin to ask for more structured student feedback in the form of a completed lab worksheet. If a metric bathroom scale is not available, skip this activity.

Vocabulary: Kilogram

2.5 USING A THERMOMETER

Discuss the need to take good care of thermometers; they are fragile. Ask students to carry them parallel to their bodies. If you use long thermometers, ask that they be held with one hand at all times. This will avoid spilling dishes and breaking thermometers.

2.6 CALIBRATING THERMOMETERS

A problem with calibrating thermometers is the altitude of the location where water boils. As one travels to higher altitudes, the temperature of boiling water drops below 100°C, because of lowered atmospheric pressure. At 760 mm of pressure (sea level), the boiling point is 100°C exactly.

Barometric Pressure mm	Boiling Temperature °Celsius
700	97.71
705	97.91
710	98.106
715	98.30
720	98.493
725	98.686
730	98.877
735	99.067
740	99.255
745	99.443
750	99.630
755	99.815
760	100.00
765	100.184

As you notice, by increasing pressure, one increases the boiling point of water. Car radiators use pressure caps for this reason. Conversely, if you go to higher elevations, you may find it impossible to cook certain foods. For example, cornmeal needs 100°C to break down its fibers. After hours of boiling, the cornmeal would still be raw. One way to correct this problem is to use a pressure cooker.

If a thermometer reads 105°C when water boils, it means that every measurement is 5° too large. Simply subtract 5 from each measurement to have calibrated values.

Vocabulary: Thermometer, calibration

2.7 FAHRENHEIT - CELSIUS CONVERSION

Give students several problems as a drill. Welcome calculators. Specify how many decimal places you wish and if you want these rounded off. Immediately following it, you will find a handy conversion table.

2.8 DATA TABLES AND GRAPHING

Go directly to the worksheet with guided practice on graphing, and follow its script. If students miss a step, they can follow instructions on that page. For clarity, the illustrations show each step in the process.

You may wish to prepare graph paper for projection. Take a sheet of 10-by-10-inch clear acetate and draw on it horizontal parallel lines, each 1 cm below the previous one. Repeat the same with parallel vertical lines, each spaced 1 cm from the other. You will have a permanent graph sheet. Cover graph grids with another sheet of acetate and tape the edges. Now you can erase without losing the grids.

The entire point of graphing is to provide a visual presentation of data. Using a graph, one can spot relationships and trends that may be difficult to perceive by gazing at numbers.

It is important that students (upper grades) learn to obtain data at points between the plotted values. They also need to project values beyond the ends of the curve. The first process is *interpolation,* the second *extrapolation.* These skills are essential. Example: What are the temperatures at 10 and 30 minutes?

Answer: The water will stop cooling when it reaches room temperature. Students need to know the room temperature. Then they can tell you the exact temperature when the cooling of the water ends.

Ask for student work and grade immediately.

Vocabulary: Axis, plot, graph, range, horizontal, vertical, variable

2.9 A GUIDED PRACTICE IN GRAPHING

You can either read this script, or have your students follow it.

2.10 LITERS AND GRADUATED CYLINDERS

Vocabulary: Graduated cylinder

2.11 VOLUME OF A REGULAR SOLID

Review with students that 1 ml = 1 cm^3 (or 1 cc in medicine.) Percentage of difference is an optional activity. The formula for percentage is: part divided by the whole multiplied by 100 ($P/W \times 100$). P represents the value which we investigate, W represents the reference value in all situations.

Example: Measured volume 37 ml, computed volume 45 ml. 37/45 × 100 = 82.2%; 100% − 82.2% = 17.8% percent of difference.

Answers to Questions:

1. Same or different.
2. Same or different.
3. Precision of measurement. Losing water while pouring; measuring the carton and including the thickness of the paper, etc.
4. The mathematical method.
5. 540,000 cm³; there are 1000 cm³ (milliliters) in a liter. 540,000 divided by 1000 (move decimal three places to the left) gives 540 liters. 540 × .45 = $243.

Students must have positive experiences with this investigation, so allow them to repeat this experiment. Help them with the math.

Vocabulary: Volume

2.12 METRIC BALANCES

Give students many opportunities to weigh small objects. Mention erasers, necklaces, watches, rings, bracelets, pieces of chalk, pebbles, straws, for example. Support students by helping them read the scales on the balances and by showing how to add up the various masses. It will be exciting when students discover that balances have an upper limit on how much mass they can measure. If time permits, allow students to measure the additional objects they have listed on their papers, since they need many experiences in weighing.

Make certain that all students have a turn at weighing. I use abandoned padlocks from a local health spa. I require that all students weigh one padlock at a time. Then they must go and exchange it for another one, in the back of the room. Any freed balance becomes available to all other students. It is good practice for future activity stations. They learn to take their data sheet with them.

Answers to #2: Portly Jim steps on two scales. The animals weigh in by being on top of a large board, supported by all the bathroom scales. The scales support the board. It is so placed as to leave the measurement windows visible. You obtain the total mass by adding the readings on all the balances. The mass of the board is subtracted to get the final mass of the horse, cow, elephant, or Sherman tank.

Vocabulary: Calibration, mass, weight

2.13 AVERAGE AND PERCENTAGE

Discuss whether you want values rounded off. Review rounding off numbers. The formula for percent is: part divided by the whole times 100 ($P/_W \times 100 = \%$).

Vocabulary: average, percent

2.14 AVERAGE MARBLES

Give lots of personal help with data.

2.15 SOLIDS DISPLACE LIQUIDS

One objective is that students learn to measure volumes of irregular solids, by using the water immersion method. Another objective is to reinforce the use of average values. You can add a few more objects of your choice and replace my suggested objects. The problem leading to endless errors is to try to do all the investigations at once. Spread this investigation over two or three days. Make sure that students have completed part 1 with marbles before going ahead. You can avoid a second problem if you focus students on each line of the data sheet. In the last column, one needs the volume of only one marble. Repeat this fact many times.

Vocabulary: displacement, meniscus

2.16 THE METRIC SYSTEM

This is a two-to-three day project. I recommend that you do it in small increments. The metric prefixes derive from both Latin and Greek words.

Distance From the Moon: 1 mile × 5280 ft/mile × 12 in/ft = 63,360 inches
250,000 miles × 63,360 inches/mile = 15,840,000,000 inches

Whole Metric System

kilometer	km	kiloliter	kl	kilogram	kg
hectometer	hm	hectoliter	hl	hectogram	hg
decameter	dam	decaliter	dam	decagram	dam
meter	m	liter	l	gram	g
decimeter	dm	deciliter	dl	decigram	dg
centimeter	cm	centiliter	cl	centigram	cg
millimeter	mm	milliliter	ml	milligram	mg

Students must have auditory experiences. Copy the step chart from the book on a board or a poster and have auditory sessions daily. Students should sing out the metric system. They must memorize the metric steps.

2.17 METRIC CONVERSIONS

Before trying any metric conversions, students must know where the decimal point is. Decimal points belong to the right of whole numbers, but most people do not write them out. Next you must stress that one moves only the point and zeroes are added as place holders only when needed. The entire process consists of moving the decimal points only. Demand that all students write the decimal point in all conversion problems, even when whole numbers are the answers.

Conversion Answers: You can use any math book for more exercises. It is good practice to reinforce this skill with small drills, twice a week for the remainder of the school year.

3700.cm;   11400.m;   .8943 dl;   14.586 kg;   63.9dl;   5000.g;   1.7 dam;
———▶3   ———▶3   ◀———2   ◀———6   ———▶1   ———▶3   ◀———1
43,000,000.mg;
———▶6

2.1 THE CENTIMETER

Purpose: The purpose of this investigation is to become familiar with metric units of measurement for distance.

Information: The *centimeter* is a metric unit used to measure distance. All nations, except the United States, use the metric system as a common standard for measurements. In 1791, the French Academy of Science introduced a decimal metric system to provide a common standard for measurement. In 1875, following a Metric Treaty in Paris, an International Bureau of Weights and Measures was established. The Metric System, which uses 10 as its base number, is easy for anyone who is familiar with the American dollar or other decimal currency. The centimeter (abbreviated *cm*) is 1/100 of a meter (abbreviated *m*). The meter is 1/10,000,000 of the distance between the earth's north and south poles. Each centimeter is subdivided into ten smaller parts, millimeters.

Equipment: 30-cm metric ruler

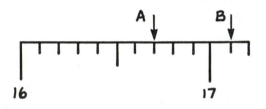

Examples: The above examples should read (A) 16.7 cm and (B) 17.1 cm. Naturally, you should round off any reading to the nearest whole millimeter.

Procedure: Measure and record the following:

Object measured	Measurement
Pencil	
Book width	
Book length	
Book height	
Shoe length	
Index finger width at knuckle	

2.2 GETTING YOUR HANDS ON A CENTIMETER

Purpose: The purpose of this investigation is to use the centimeter.

Information: Ten millimeters (mm) make up 1 centimeter (cm). Ten centimeters make up 1 decimeter (dm) and 10 dm make 1 meter (m). One meter has 10 decimeters, 100 centimeters, or 1000 millimeters. One dollar has ten *dimes* as 1 meter has 10 *decimeters,* and one dollar has 100 *cents* as 1 meter has 100 *centimeters.*

Equipment: Metric ruler, this sheet

Procedure 1:

1. Shown below are many lines. Measure them in centimeters.
2. Letter each line of a clean page with a letter of the alphabet, starting at *A* and ending at *Z*.
3. Next to the letter place the measurement of the line on this sheet. Remember to label each value as a centimeter.

Sample Data Setup: (This is for data setup only; it is *not* actually correct!)
Measurement in cm
A   17.5 cm
B   27.9 cm

A. _____

B. _____

C. _____

D. _____

E. _____

F. _____

G. _____

H. _____

I. _____

J. _____

K. _____

L. _____

M. _____

N. _____

O. _____

P. _____

Q. _____

R. __

S. ___

T. ____

U. _____

V. _____

W. _____

X. _

Y. _____

Z. _____

Procedure 2: Change your measurements into millimeters, by moving the decimal one place to the right. Examples: 14.3 cm = 143. mm; 25.4 cm = 254. mm; 0.7 cm = 7. mm. Pay special attention to where the decimal points go. Sometimes you will have a whole number ending with a decimal point. Leave the decimal point there. Your page should look like the following:

	Measurement in cm	Measurement in mm
A	17.5 cm	175. mm
B	27.9 cm	279. mm

2.3 I AM A METRIC CREATURE

(Partner activity)

Purpose: The purpose of this activity is to use metric units.

Information: Measurements are part of daily life. You need to become familiar and comfortable with using metric units.

Equipment: One meter stick for every student, data sheet prepared in advance.

Procedure: Prepare a data sheet listing on the left the names of objects to measure. Start with personal measurements:

1. Width of smile
2. Length of index finger
3. Distance of one step
4. Length of shoe
5. Width of wrist
6. The lengths of tapes *A, B, C, D,* and *E.*
7. Door width and height
8. Room width and length.

Measure all the items listed on your data sheet. When finished, return to your seat and convert all measurements into millimeters.

2.4 MY HEIGHT AND MASS

(Partner activity)

Purpose: The purpose of this investigation is to use metric units.

Information: With this activity, you begin a long-term investigation. You will measure your own height and mass for two months. At the end of this period, you will graph the data and present them to your teacher as a laboratory report.

Equipment: Two meter rulers fastened to the wall, (measurement station), one separate ruler, one metric bathroom scale.

Procedure:

1. Every day, or at least three times a week, measure your own height with the help of another student.
2. Immediately record the measured value on the data sheet.
3. Measure your own mass in kilograms and immediately record it. Plan carefully to use the measurement station and the bathroom scale when other students are not using them.

Sample Data Setup:

DATE	MY MASS	MY HEIGHT
9-29-	38.5 kg	112.7 cm
9-30-	38.5 kg	112.4 cm
etc.	etc.	etc.

Questions: Answer all questions in full.

1. How much did you grow in the last eight weeks?

_____

_____

2. Did your body mass change? By how much?

_____

_____

3. Did you notice that your measurements varied? Why do you think this happened?

_____

_____

4. If you were to repeat this investigation, how would you do it differently next time?

_____

_____

5. Explain why you would make these changes.

_____

_____

2.5 USING A THERMOMETER

Purpose: The purpose of this investigation is to practice using a thermometer.

Information: In the metric system, Celsius thermometers use the freezing of water at 0° C and the boiling of water at 100° C as reference points. There are 100 degrees between freezing and boiling of water. In some countries, the Celsius thermometer is known as a *centigrade* thermometer. Human body temperature is 37°C (98.6° Fahrenheit), 22°C is a comfortable room temperature (72°F), and 37.7°C is the same as 100°F, quite hot.

Equipment: Cup, thermometer, warm water

Procedure:

1. Pour warm water into cup.
2. Place thermometer in cup.
3. Record the temperature of water every minute for fifteen minutes.

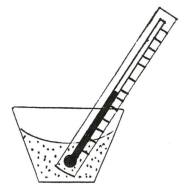

Sample Data Set-Up: Prepare a sheet of paper:

TIME	TEMPERATURE
0	_____ °C
1 min.	_____ °C
2 min.	_____ °C
...	
14 min.	_____ °C
15 min.	_____ °C

Questions:

1. Were all temperature readings exactly on the marks?

_____

_____

2. If they were not, could you estimate the tenths between the marks?

_____

_____

Name _____ Date _____ Period _____

2.6 CALIBRATING THERMOMETERS

(Partner activity)

Purpose: The purpose of this investigation is to learn to calibrate a thermometer.

Information: A thermometer is usually a glass tube filled with a liquid, which expands with heat. Thermometers are filled with alcohol, mercury (a metal which is liquid at room temperature), or sometimes a gas. A common problem with most inexpensive lab thermometers is that they show an incorrect temperature. Calibration is the process of adjusting the value of your thermometer reading to that of a universal standard.

Water boils at 100°C (212°F).   Water freezes at 0°C (32°F).

Equipment: Beaker, 250 ml water, hot plate, student thermometer

Procedure:

1. Place your thermometer into beaker with boiling water.
2. Keep it in there for about a minute or until the indicator liquid stops moving up.
3. Record the temperature of boiling water, exactly as the thermometer shows it.

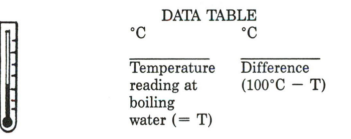

DATA TABLE

°C	°C
_____	_____
Temperature reading at boiling water (= T)	Difference (100°C − T)

Example: If your thermometer reads 87°C, then 100 − 87 = 13°C. Your thermometer reads 13° less than it should. All your measurements are lower by 13° than what they should be. You correct this problem by simply adding 13° to every reading. You have just converted your inexpensive thermometer into a precision model. If your thermometer reads above 100°C, when water boils, subtract the difference from your measurements.

Questions:

1. If there is a difference between 100°C and your thermometer reading, how do you use this information to adjust your readings?

_____

_____

2. The difference is 5°C because your thermometer shows water boiling at 95°C. With the same thermometer you measure a liquid at 27°C. What should your adjusted reading be for the real temperature of the liquid?

_____

_____

3. What should the 27°C value be if, in calibrating, your thermometer showed water boiling at 105°C?

_____

2.7 FAHRENHEIT–CELSIUS CONVERSION

Purpose: The purpose of this investigation is to learn to convert temperatures both ways between the Celsius and Fahrenheit scales.

Information: Thermometers are devices used to measure temperature. Most of the world uses the Celsius scale, but the U.S.A. uses the Fahrenheit scale.

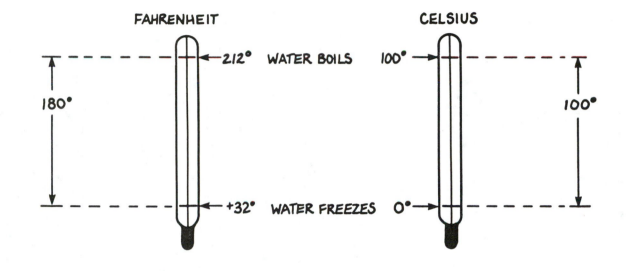

Information: One notices that a Celsius thermometer has 100 degrees between boiling and freezing, while a Fahrenheit one has 180 degrees. So by dividing 180 by 100 we find out that 1°C = 1.8°F. Next, if we account for the +32°, then we arrive at the following conversion formulas:

Fahrenheit to Celsius:

$$C° = \frac{(F° - 32)}{1.8}$$

Example: 88°F = X°C

(88 −32)/1.8 = 31.1°C

Celsius to Fahrenheit:

$$F° = (C° \times 1.8) + 32$$

Example: 40°C. = X°F

(40 × 1.8) + 32 = 104°F

FAHRENHEIT-CELSIUS CONVERSION TABLE

F°	C°	F°	C°	F°	C°	F°	C°	F°	C°
1 =	−17.22	51 =	10.56	101 =	38.33	151 =	66.11	201 =	93.89
2 =	−16.67	52 =	11.11	102 =	38.89	152 =	66.67	202 =	94.44
3 =	−16.11	53 =	11.67	103 =	39.44	153 =	67.22	203 =	95.00
4 =	−15.66	54 =	12.22	104 =	40.00	154 =	67.78	204 =	95.56
5 =	−15.00	55 =	12.78	105 =	40.56	155 =	68.33	205 =	96.11
6 =	−14.44	56 =	13.33	106 =	41.11	156 =	68.89	206 =	96.67
7 =	−13.89	57 =	13.89	107 =	41.67	157 =	69.44	207 =	97.22
8 =	−13.33	58 =	14.44	108 =	42.22	158 =	70.00	208 =	97.78
9 =	−12.78	59 =	15.00	109 =	42.78	159 =	70.56	209 =	98.33
10 =	−12.22	60 =	15.56	110 =	43.33	160 =	71.11	210 =	98.89
11 =	−11.67	61 =	16.11	111 =	43.89	161 =	71.67	211 =	99.44
12 =	−11.11	62 =	16.67	112 =	44.44	162 =	72.22	212 =	100.00
13 =	−10.56	63 =	17.22	113 =	45.00	163 =	72.78	213 =	100.56
14 =	−10.00	64 =	17.78	114 =	45.56	164 =	73.33	214 =	101.11
15 =	− 9.44	65 =	18.33	115 =	46.11	165 =	73.89	215 =	101.67
16 =	− 8.89	66 =	18.89	116 =	46.67	166 =	74.44	216 =	102.22
17 =	− 8.33	67 =	19.44	117 =	47.22	167 =	75.00	217 =	102.78
18 =	− 7.78	68 =	20.00	118 =	47.78	168 =	75.56	218 =	103.33
19 =	− 7.22	69 =	20.56	119 =	48.33	169 =	76.11	219 =	103.89
20 =	− 6.67	70 =	21.11	120 =	48.89	170 =	76.67	220 =	104.44
21 =	− 6.11	71 =	21.67	121 =	49.44	171 =	77.22	221 =	105.00
22 =	− 5.56	72 =	22.22	122 =	50.00	172 =	77.78	222 =	105.56
23 =	− 5.00	73 =	22.78	123 =	50.56	173 =	78.33	223 =	106.11
24 =	− 4.44	74 =	23.33	124 =	51.11	174 =	78.89	224 =	106.67
25 =	− 3.89	75 =	23.89	125 =	51.67	175 =	79.44	225 =	107.22
26 =	− 3.33	76 =	24.44	126 =	52.22	176 =	80.00	226 =	107.78
27 =	− 2.78	77 =	25.00	127 =	52.78	177 =	80.56	227 =	108.33
28 =	− 2.22	78 =	25.56	128 =	53.33	178 =	81.11	228 =	108.89
29 =	− 1.67	79 =	26.11	129 =	53.89	179 =	81.67	229 =	109.44
30 =	− 1.11	80 =	26.67	130 =	54.44	180 =	82.22	230 =	110.00
31 =	− 0.56	81 =	27.22	131 =	55.00	181 =	82.78	231 =	110.56
32 =	0.00	82 =	27.78	132 =	55.56	182 =	83.33	232 =	111.11
33 =	.56	83 =	28.33	133 =	56.11	183 =	83.89	233 =	111.67
34 =	1.11	84 =	28.89	134 =	56.67	184 =	84.44	234 =	112.22
35 =	1.67	85 =	29.44	135 =	57.22	185 =	85.00	235 =	112.78
36 =	2.22	86 =	30.00	136 =	57.78	186 =	85.56	236 =	113.33
37 =	2.78	87 =	30.56	137 =	58.33	187 =	86.11	237 =	113.89
38 =	3.33	88 =	31.11	138 =	58.89	188 =	86.67	238 =	114.44
39 =	3.89	89 =	31.67	139 =	59.44	189 =	87.22	239 =	115.00
40 =	4.44	90 =	32.22	140 =	60.00	190 =	87.78	240 =	115.56
41 =	5.00	91 =	32.78	141 =	60.56	191 =	88.33	241 =	116.11
42 =	5.56	92 =	33.33	142 =	61.11	192 =	88.89	242 =	116.67
43 =	6.11	93 =	33.89	143 =	61.67	193 =	89.44	243 =	117.22
44 =	6.67	94 =	34.44	144 =	62.22	194 =	90.00	244 =	117.78
45 =	7.22	95 =	35.00	145 =	62.78	195 =	90.56	245 =	118.33
46 =	7.78	96 =	35.56	146 =	63.33	196 =	91.11	246 =	118.89
47 =	8.33	97 =	36.11	147 =	63.89	197 =	91.67	247 =	119.44
48 =	8.89	98 =	36.67	148 =	64.44	198 =	92.22	248 =	120.00
49 =	9.44	99 =	37.22	149 =	65.00	199 =	92.78	249 =	120.56
50 =	10.00	100 =	37.78	150 =	65.56	200 =	93.33	250 =	121.11

Name _____ Date _____ Period _____

2.8 DATA TABLES AND GRAPHING

Purpose: The purpose of this investigation is to learn and review the process of preparing x-y graphs and data tables.

Information: *Graphing* is a way of representing data (numerical values) in picture form. The principal types of graphs are: vertical bar, horizontal bar, pie, and x-y.

Procedure: Graph the values provided in *Sample Data.* These are obtained by allowing a pan of hot water to cool over time. The temperature measurements were taken every three minutes.

Equipment: One piece of graph paper (¼ in squares), ruler, sharp #2 pencil.

Sample Data Setup:

WATER TEMPERATURE LOSS IN DEGREES CELSIUS

Time	Temperature Celsius
START 0	90°C
3 min.	86°C
6 min.	82°C
9 min.	78°C
12 min.	74°C

Time	Temperature Celsius
15 min.	71°C
18 min.	68°C
21 min.	65°C
24 min.	63°C
27 min.	62°C

Information: Use the horizontal axis for plotting times and the vertical axis for plotting temperatures. Prepare a graph of the above data. Your temperature starts at 90°C and ends at 62°C, giving you a range of 28 degrees. *Range* is the spread between the maximum and minimum values. The time starts at 0 and goes on for 27 minutes. The range is 27 minutes. When you decide on how many squares to allow for each *variable,* time and temperature, you need to know their ranges. A variable is a value which changes during an experiment. There are two types of variables: *dependent* and *independent.* The independent variable is the one you control during the experiment. Here, it is the time. You plot independent variables on the *X*-axis (the horizontal line). A dependent variable is one which changes as you control the independent one. Here, the dependent variable is the temperature of water, which cools over time. Your body height is a variable dependent on your age (an independent variable). You plot the dependent variable on the vertical axis (*X* = axis).

Questions:

1. What was the water temperature at 5, 7, and 10 minutes?

_____

2. What was the temperature change over a period of ten minutes?

_____

3. What was the temperature change from 5 to 15 minutes?

_____

4. What is the cooling rate of water per minute at 1 minute and at 27 minutes?

_____

5. How long will it take before the cooling stops?

_____

6. Is there any specific information, not part of this experiment, which you need to accurately answer questions?

_____

_____

_____

2.9 GUIDED PRACTICE IN GRAPHING

Purpose: The purpose of this activity is to provide a script for graph preparation.

Procedure: Place your graph paper in front of you so that the long side of the page runs from your left to your right. The page should have about 44 squares horizontally and about 32 vertically. Count them and mark your numbers in a corner.

1. Locate the lower left corner of the paper.

2. Count six lines to the right, then five lines up and mark the spot with a dot.

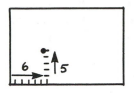

3. Draw one thick line across the entire paper from left to right through this dot, covering the graph line. Label it *X*-axis, right above the line, near the left edge.

4. Draw a thick line up and down the paper through this same dot, covering the graph line. Mark it as the *Y* axis. Label it below the line.

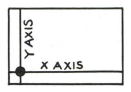

5. You should have about 28 squares left vertically and about 38 horizontally. This allows you to use one square for each minute horizontally and one degree per square vertically. Mark each vertical line below the *X*-axis just one square down from the line. Continue all the way to the right of the starting point. When finished mark a horizontal line, one square side, to the left of the *Y*-axis.

6. Label each mark below the horizontal line, starting with the vertical axis as 0, then continuing to the end of the paper. The number must be just below each line.

7. Label each horizontal line, starting with the highest horizontal line as 90, then 89, 88, 87, 86, etc. Continue until you run out of marks. Do include the horizontal axis as one mark.

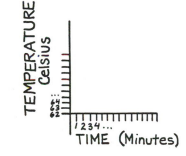

8. Label below the horizontal numbers, in large letters: *TIME (minutes)*.

9. Turn the page so the Y axis line is from your left to your right, on top of the page.

10. Label above the numbers in large letters: *TEMPERATURE (degrees Celsius)*.

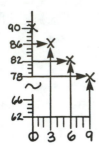

11. Now plot all the values. Your first value is at time = 0, so you use the vertical reference axis and locate 90°. At the point where the 0 and the 90 lines cross, place a small *x*.

12. Follow up the 3-minute line and look where it crosses with the 86° line. Mark it with a small *x*.

13. Continue until you use all the data in table.

14. Using a ruler, connect all the centers of the *x* marks. This line is called a *curve,* even if sometimes it looks like straight line segments joined together.

15. Above the curve write *cooling rate of water.*

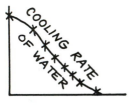

16. Place your name, period, date, and room number in the upper right corner of the paper.

2.10 LITERS AND GRADUATED CYLINDERS

Purpose: The purpose of this investigation is to introduce the liter and the use of graduated cylinders.

Information: In the metric system, the *liter* is the unit for measuring liquids. In practical terms, the liter is slightly larger than the quart. If a quart were to be represented by $1.00, then a liter would be $1.05, or a nickel more. A liter is 5 percent more than a quart.

Procedure: Today you will be an artist! You will carve out of clay a small cube, 1 cm long on every edge.

Information: A *liter* is the volume occupied by 1000 cubic centimeters (cm³), (cc to your doctor). Since there are 1000 cc in a liter, then 1 cm³ = 1 milliliter (1/1000 liter). A liter box is a cube 10 cm × 10 cm × 10 cm = 1000 cubic centimeters. It is important to note that 1 cm³ of water (1 ml) weighs exactly 1 gram. One liter of water weighs exactly 1 kg. Now you know how mass, bulk, and distance relate in the metric system.

Special Information: Reading graduated cylinders is tricky, for water sticks to the edges of dishes and curves upwards. This curve is called the *meniscus*. Therefore, as you look at the graduated cylinder, you will see two lines; always read the lower one.

Equipment: Clay or florist's clay, cardboard, dull knife, shallow pan, 100 ml graduated cylinder, jar, food coloring, water.

Procedure: Observe the graduated cylinder and you will notice marks. Usually it has a 0 at the bottom and 100 ml at the top. Fill the graduated cylinder to 12 ml, 25 ml, 41 ml, 63 ml, 88 ml, 99 ml. Now fill the graduated cylinder with water, pour out a little and read. Repeat three or four more times. Record your readings and write in the units (milliliters). Example: 17 ml, 21 ml, etc.

2.11 VOLUME OF A REGULAR SOLID

(Partner activity)

Purpose: The purpose of this investigation is to compare the volume measured during an activity with one calculated mathematically.

Information: You measure volumes in several ways:

A. Using the tools of mathematics and geometry
B. Using liquid measures

To measure the volume of a solid like a shoe-box, a parallelepiped, you need to multiply its length by its width and by its height. To restate this in a brief mathematical way. $V = L \times W \times H$. If a box measures 2 cm $\times$ 3 cm $\times$ 4 cm, then $V = 2 \times 3 \times 4 = 24$ cm³, which you would read as 24 *cubic centimeters.*

VOLUME IS ALWAYS EXPRESSED IN CUBIC UNITS!

Equipment: Two relatively waterproof paper cartons, one small and one larger, graduated cylinder, metric ruler, water

Procedure:

1. Measure inside the small carton on all four sides 5 cm from the bottom.
2. Mark this height by drawing a line around the carton and enter this value in the data sheet under height.
3. Measure both the length and width of the carton and enter measurements under length and width.
4. Repeat the entire procedure, this time using the larger carton.
5. Calculate the volumes of the cartons by using the mathematical method.
6. Enter the volumes in the boxes under calculated volume.
7. Fill each container with water to your pencil mark and measure the water using the graduated cylinder. Record this value under measured value.

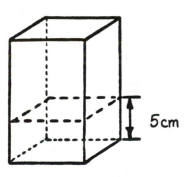

Help: Remember that 1 ml = 1 cm³

Sample Data Setup:

carton	length cm	width cm	height cm	calculated volume cm³	measured volume cm³	difference volume cm³	% diff.
small							
large							

Help: To figure out percentages of difference (optional), divide measured volume by the calculated volume and multiply the answer by 100. Subtract this value from 100 percent, and the result is your answer.

Questions:

1. Looking at the smaller carton, how does the calculated volume compare with the measured volume?

   _____

2. What is the difference between the volumes in the larger carton?

   _____

3. Why do you think there is a difference?

   _____

4. Which method do you consider as more reliable? Why?

   _____

5. Figure out the volume of a larger box, measuring 50 cm × 90 cm × 120 cm. Recall that 1000 ml = 1 l. If you fill the box with milk, how many liters of milk do you need? If milk costs $0.45 per liter, what is the cost to fill the box? _____

2.12 METRIC BALANCES

(You will work in a small group.)

Purpose: The purpose of this investigation is to use a metric balance to measure the masses of several objects.

Information: The metric unit of mass is a gram (g). The Science Academy in Paris defines a gram as a mass which is represented by a cube of water with sides 1 cm each. You can imagine this tiny mass when you think of the mass of a medium-size paper clip: not much to it. It would take about 29 of these paper clips to make up one ounce and 454 paper clips to equal one pound. Since the gram is so small, people around the world use mostly the kilogram (1 kg = 1000 g) as a convenient unit of measurement. Pharmacists and scientists use milligrams (mg = 1/1000 g). You can probably recall reading milligrams on vitamin packages. Weight, a force, is the pull of earth on all material objects. In outer space there is no gravity and therefore no weight. There is only mass, the property of being a material object. On earth, when you throw a ball up, it falls down. In outer space a ball floats around.

Questions:

1. What will happen in outer space, if you throw a ball up?

_____

_____

Calibration: To provide accurate readings, all balances must be calibrated. To calibrate means that the pointer must point to zero with all the masses set at zero. This happens with no object in the measuring dish. Balances usually have a screw or other device to calibrate them.

On triple beam balances, one weighs objects by sliding masses to the right. It is most important that the masses sit in the notches provided, not between them. Some balances have dials to replace the smaller masses.

Procedure 1: Measure the mass of several small objects. Prepare a data sheet similar to the sample one. Since you wish to enjoy this activity, have some fun and make some guesses on what these objects may weigh. In science this guess is an *estimate*. It does not matter if you are off, because it is only an estimate. As you become more experienced, your estimates will get closer to the measured values.

Sample Data Setup:

OBJECT MEASURED	ESTIMATED GRAMS	MASS (g)
pencil	60 g	14.7 g
ring	20 g	15.1 g
etc.	? g	? g

Question:

1. What objects would you like to weigh, if given the opportunity?

_____

_____

2. Make a list below your data page as if you were going to weigh these additional objects.

Bonus:

Procedure 2:   Invent a strategy (with your group) for how to weigh a person weighing over 200 kg (440 lb) and a live horse or cow (no cutting to pieces, please). The animal has an approximate mass of 550 kg (1200 lb). You have six bathroom scales which can weigh up to 150 kg (330 lb). Explain on the back of your data sheet how you would do it and draw a picture to illustrate!

2.13 AVERAGE AND PERCENTAGE

(Partner activity)

Purpose: The purpose of this activity is to use average values in science and to introduce percentages.

Information: When measuring mass, volume, or distance, there is always a chance for error in measurements. To reduce this possibility, you measure several times and average the values. The *average value* is far more reliable than any one measurement.

Equipment: Padlocks or other objects, balance

Procedure:

1. Find the mass of each object.
2. Repeat measurements for each different object.
3. Repeat until your data table is filled in.
4. Do not measure one object five times in a row: alternate.
5. Calculate averages.
6. (Optional) Round off values to one decimal place.

Help With Average:

1. You measured the mass of a marble as 2 g, 2.2 g, 2.3 g, 2.9 g and 2.5 g.
2. The average value is found by adding the five values, $2 + 2.2 + 2.3 + 2.9 + 2.5 = 11.9$
3. Next you divide 11.9 by 5 (since you measured five times) Voilà, the average value is 2.38 g.
4. Round off this value to 2.4 since .38 is closer to .40 than .30.

Sample Data Setup:

OBJECT #	mass g.	mass g.	mass g.	mass g.	mass g.	AVERAGE g.

© 1991 by The Center for Applied Research in Education.

Question: What is the average mass for each object measured?

_____

_____

Information: Percent (%) tells us how many times an event does or does not work out of 100 times. *Percent* means *out of 100*. When we measure we need to know sometimes what is the percent of a measurement compared to the average value. We like to know the percent of the value 2. g compared to the average of 2.4 g. Perform as follows: $2/2.4 \times 100 = 83.33\%$. You take the measurement under question, divide it by the average and multiply it by 100. This gives you the value of the percent of closeness of that measurement to the average. Be prepared at times to have a percent greater than 100. $2.9/2.4 \times 100 = 120.8\%$. This tells us that your measurement was 20.8% greater than the average. A large percentage away from the average value tells you that the measuring was careless.

Question:

1. Take one object and compare its lowest measured value to the average value, as in the example. Show your work.

2.14 AVERAGE MARBLES

(Partner activity)

Purpose: The purpose of this investigation is to do an activity which uses averages.

Information: Marbles occupy space in air. When placed on a balance you can measure their *mass.* The property of having weight is common to all material objects subject to the force of gravity. *Weight* is measured using the *gram* as the unit. A gram is the weight of one cubic centimeter of water. Since one cubic centimeter conveniently equals one milliliter, one gets the weight of water directly, by measuring its volume with a graduated cylinder. The number of milliliters of water equals the number of grams. (This is one major advantage of using the metric system.)

Equipment: Metric scale, (weights where needed), ten marbles, holding dish.

Procedure:

1. Calibrate your balance.
2. Measure the mass of the dish. Record it.
3. Place one marble in the dish. Measure the combined masses of the dish and the marble. Record this mass.
4. Place a second marble in the dish. Repeat the measurement and record it.
5. Use the same procedure until you have measured ten marbles in the dish.

© 1991 by The Center for Applied Research in Education.

Sample Data Setup:

OBJECTS	mass (g) total	mass of ALL marbles (g)	AVERAGE mass of ONE marble only (g)
DISH		XXXXXXXXXXXXX	XXXXXXXXXXXXXXX
+1 MARBLE			
+2 MARBLES			
+3 MARBLES			
+4 MARBLES			
+5 MARBLES			
+6 MARBLES			
+7 MARBLES			
+8 MARBLES			
+9 MARBLES			
+10 MARBLES			

(Now average the entire column of marble averages) AVERAGE _____ g

Help: Data table: to learn the mass of marbles only, subtract from their masses the mass of the dish. To obtain the mass of only one marble, divide the mass of several marbles by the number of marbles in that group. Remember at the very end to average all averages. To average two numbers, you add them together and divide by 2. If you average 7 and 9, the sum is 16 and the average is 8. (16 divided by 2).

Questions:

1. What is the mass of a single marble? _____

2. Why should one make many measurements and then average them?

_____

_____

2.15 SOLIDS DISPLACE LIQUIDS

(Partner activity)

Purpose: The purpose of this investigation is to measure the volume of irregular solids and to practice averages.

Equipment: Small jar, marbles, shallow pan.

PART ONE

Procedure:

1. Place a small jar in a flat pan and fill jar with water until it is full.
2. Gently place marbles in jar until the water overflows.

Question:

1. Why did the water overflow?

_____

_____

Information: When you place any material object in a container of water, the object takes room and displaces (moves out of the way) its own volume of water. When you enter a tub, the water level rises. When you make hard-boiled eggs, you are careful not to fill the pot with too much water, because every time you place an egg in the pot, the water level rises.

PART TWO

Equipment: One graduated cylinder 100 ml, 10 marbles, colored water, 15 pennies, 60 paper clips, 3 large metal nuts or 1 oz fishing lead sinkers, shallow dish, pieces of string or sewing thread, jar.

Procedure 1:

1. Place about 60 ml of water in the graduated cylinder and record the exact level on the data sheet. The water level curves downward from the sides of the jar; this surface curvature is the *meniscus.* Always read the water at the lowest level of the meniscus. This is your initial *0* level.
2. Drop a marble in the jar and note the rise in the water level. Record this new level as your *1* (marble) level. Use the same size marbles for this entire activity.
3. Record your data in a data table like the one provided.
4. Construct a bar graph from the data. Use the vertical axis for volume (ml), the horizontal axis for the number of marbles.
5. If you need more materials than are in your kit, borrow from another team.

Sample Data Setup:

MARBLES

number of marbles	measured meniscus level (ml)	average volume of one marble (ml)
0		XXX
1		
2		
3		
4		
5		
6		
7		
8		
9		
10		

Average value: _____ ml

Strategy: Assume that your cylinder has 60 ml of water. *This is your reference point.* When you drop one marble, it goes up to 64 ml. You say, therefore, that one marble has the volume of 4 ml. (64 − 60 = 4). (This is only an example!) Now, if you have three marbles in the water and the reading is 72 ml, then 72 − 60 = 12 ml. This means that three marbles occupy 12 ml. Divide 12 by 3 to find out that each marble is 4 ml. Once you find the volume of one marble many times, average the values to improve accuracy.

Procedure 2: Do the same investigation with the pennies.

PENNIES

number of pennies	measured meniscus level (ml)	average volume (ml) one penny
0		XXX
15		
20		
25		
30		
35		
40		
45		
50		
55		
60		

Average value: _____ ml

Procedure 3: Do the same investigation with the paper clips.

PAPER CLIPS

number of paper clips	measured meniscus level (ml)	average volume (ml) one paper clip
0		XXX
15		
30		
60		
15		
30		
60		

Average value: _____ ml

Procedure 4: Do the same investigation with metal nuts or lead sinkers.

LEAD SINKERS OR STEEL NUTS
(use same one many times)

number of nuts	measured meniscus level (ml)	average volume (ml) one nut
0		XXX
1		
2		
3		
1		
2		
3		

Average value: _____ ml

Questions:

1. How many pennies equal the volume of one marble? How many paper clips equal the volume of one marble?

   _____

   _____

2. How many pennies/paper clips would equal the volume of one metal nut.

   _____

   _____

3. What is the volume of 20, 25, and 48 marbles?

   _____

   _____

4. What is the volume of 100 paper clips?

   _____

   _____

5. What is the volume of 100 pennies?

   _____

   _____

2.16 THE METRIC SYSTEM

Purpose: The purpose of this investigation is to understand and become competent in the basic setup of the metric system.

Procedure 1: List on a piece of paper all units of bulk, mass, and distance used in the U.S.A. Use your memory only. If you can list more than thirty without having to look them up, you are a genius! If you understand fewer than ten, look out!

Procedure 2: The following is a good example of the difficulty in changing units within the US system (FPS, Foot-Pound-Second): Convert the distance from the Earth to the Moon, 250,000 miles, to inches.

1 mile = 5280 ft.; 1 ft. = 12 in.

Information: If you give up on the US system, join the rest of the world and learn the metric system (MKS, Meter-Kilogram-Second). Before you learn anything, you must meet three important prerequisites. See if you qualify:

1. You must know your right from your left.
2. You must be able to count from one to six.
3. You must be able to memorize nine tiny words:

    kilo, hecto, deka, deci, centi, milli, meter, liter, gram.

In the metric system, it is not necessary to remember any conversion factors or do any complex arithmetic. To change from one unit into another, you move the decimal point to the right or to the left. You count the places. That is all!

Required Memorization

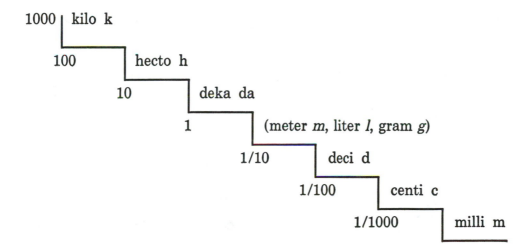

The number in front of the word tells you how much that word stands for: kilo = 1000, deka = 10, etc. The letter to the right of the word is its abbreviation. You do not need to write these words time after time, use only the abbreviation.

Procedure 3·

1. ·On a piece of paper, write down from top to bottom on the left the six words for quantity, leaving a blank line for the one, just like this:
   kilo
   hecto
   deca
   _____

   deci
   centi
   milli

2. Add the word *meter* after each prefix and you will have:
   kilometer
   hectometer
   decameter
   meter
   decimeter
   centimeter
   millimeter

You now know the entire system for meters. Next to each unit write its abbreviation. When finished, repeat the activity for liters and grams.

Name _____ Date _____ Period _____

2.17 METRIC CONVERSIONS

Purpose: The purpose of this investigation is to show the ease with which one can convert units within the metric system.

Information: Do not go on unless you have memorized the step chart in the previous activity. Go back and memorize it. Test yourself and see if you can write it down on a piece of paper, without looking at it. Only when you can do it, go on.

In the metric system one changes units very simply. To count the steps on the metric scale you must move to go to the new unit from the given one. You must notice if you go to the right or to the left. Then you move the decimal point as many places as you counted, in the same direction as you moved on the step scale. Draw a step scale and have it in front of you.

Examples: Perform the following exercises:

17.5 dam = _____ cm (Move from dam to m, dm, cm—three places to the right. So move the decimal point three places to the right. In other words, 17.5 dam = 17,500. cm).

3285 ml = _____ kl (Move from ml to cl, dl, l, dal, hl, kl—six places to the left—so move the decimal point six places to the left. 3285 ml = .003,285 kl.

Procedure: Convert the following values:

1. 3.7 m = _____ cm;
2. 11.4 km = _____ m;
3. 89.43 ml = _____ dl;
4. 14,568,000 mg = _____ kg;
5. 6.39 l = _____ dl;
6. 5 kg = _____ g;
7. 17 m = _____ dam;
8. 43 kg = _____ mg;

MATTER

TEACHER'S SECTION

In this section we will examine properties of matter:

1. Molecules in motion
2. Phases of matter
3. Evaporation and condensation
4. Variables in evaporation
5. Solutions
6. Schlieren
7. Dissolving solids
8. Supersaturated solutions-crystals
9. Osmosis
10. The law of conservation of matter
11. Gases
12. Making carbon dioxide
13. Adhesion and cohesion
14. Surface tension
15. Wetting agents
16. Soap bubbles
17. The volume of a drop of water

3.1 THE MOTION OF MOLECULES

Students must acquire a knowledge of basic terms involved with the classification of matter. During the activity, move around freely and ask students why the dye spreads out. The result of this activity is a solution, used to show the behavior of solutions of liquids in liquids. Molecular motion and collisions will explain why. Make clear that molecules are constantly in motion. Their internal level of energy determines how fast and how far they will move. The energy level also determines whether a substance is a solid, a liquid, or a gas. With an increase of heat energy, the molecular motion increases, and, with it, the speed of solution.

In this experiment students measure the spread of a drop of food coloring. The molecular motion and collisions between molecules of both liquids cause the spread. The resulting solution (mixture) is colored water.

The actual measurement is rather tricky, for each team will have a different underwater color blob. Suggest that students measure the widest point of the blob. Have the partners trade off, so both get measuring experience. Discrepancies in measurements will lead eventually to a group consensus that there is a need to set standards on how to measure. Students need to get consistent data. For the same reason, scientists publish their findings in journals. These must be repeatable with the same outcomes world-wide, if the discoveries are to be credible.

OPTIONAL: As a warm-up, have students measure segments and lines on the lab worksheet.

1. Place lab worksheet next to experiment page.
2. Have students read Purpose and Information. A student can read it aloud.
3. Point out the key definitions for *atoms, elements,* and *molecules.*
4. Have students copy these at home, adding to their vocabulary page.
5. Have students prepare a data sheet like the sample.
6. Model the experiment:
   a. Take a bowl of water.
   b. Place in it a drop of food coloring.
   c. Have the entire class look at it.
   Be alert that some color will spread on the surface, while the actual color drop will sink to the bottom of water.
7. Students measure the bottom drop.
8. Assign the completion and writing of the lab worksheet as a home activity. Class time must be spent on doing the activity.
9. Have students try to perform the activity as a first run. Give only ten minutes, then call off activity.
10. Give the real activity on the following day.

Closure on Activity:

1. Have students graph their data, using a vertical bar graph.
2. Help with conclusions; have them reread Purpose.
3. Suggest that they write the following in their conclusions:
   a. the spread values for each minute (lowest thinking skill);
   b. the differences between spread values (higher thinking skills);
   c. the significance and meaning of the spread values (highest thinking skills);
4. Ask the following:
   a. "Can you write other examples which support your position?"
   b. "How can you improve this experiment?"
   c. Does the experiment prove the hypothesis (see purpose)?

The experiment proves molecular motion, by circumstantial evidence. We cannot see molecules, for they are too small.

Solution of a liquid in a liquid is also shown. Water is the solvent (it is the larger quantity), and the food coloring is the solute.

Vocabulary: Atom, element, molecule

3.2 PHASES OF MATTER

Part of the water becomes water vapor. As closure to this activity, ask for conclusions. Some examples:

1. It takes energy to melt ice into water.
2. The part of the ice closest to the heat source melts first.
3. Smell is gas molecules which enter one's nose.
4. Heat evaporates water.
5. Additional examples:

SOLIDS	LIQUIDS	GASES
butter	melted butter	fat vapors
rock	lava	vaporized lava
ice	water	water vapor
frozen perfume	perfume	perfume vapor
frozen juice	juice	juice vapor
steel	liquid steel	steel vapor
	ammonia	ammonia vapor
	gasoline	vaporized gasoline
onion	onion juice	smell of onion
chicken	chicken soup	smell of soup
wax	liquid wax	smell of candle
wool		burned wool smell
dry ice (CO_2)		CO_2 gas (does not exist as a liquid)

Matter has really four phases. The fourth state is *plasma*. As heat energy is applied, an ice cube (solid) melts (liquid) and evaporates (gas). If heat continues to be applied to the gas, the atoms of gas become supercharged until the individual atoms break up. This means that the nuclei and the electrons break apart. The ejected free subatomic particles (protons, neutrons, and electrons) and the atomic bonding energy become the *plasma*. Plasma occurs only at temperatures around 6 million degrees Celsius, as on the surface of our sun or other stars. Plasma energy from the sun is the *solar wind*. The point in space where the solar wind ends is the *heliopause*. We do not know where this point is. NASA is investigating its precise location with the Voyager spacecraft.

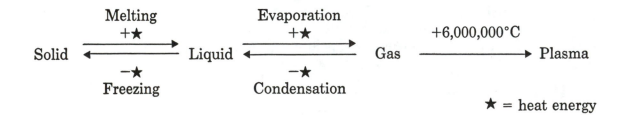

3.3 EVAPORATION AND CONDENSATION

Results: After a short while, the balance will tip and the cold water will appear to become heavier. The hot water evaporated more quickly than the cold water. There is less mass of H_2O in the hot dish.

Answers:

1. One is hot, the other is colder.
2. The cold water has its molecules closer to each other, so it appears that it is heavier compared to the same volume of warmer water. The heat from the blackboard helps the water molecules evaporate, just as body heat helps perspiration to dry. On days with high humidity, the air is saturated with water vapor. You feel sticky and uncomfortable, since perspiration evaporates at a much slower rate. This explains why one can be in a desert and feel comfortable—sweat evaporates almost instantly and you feel cool.

TEACHER ACTIVITY

Procedure: Wet the hand of each student with a drop or two of rubbing alcohol.

Questions:

1. Did you feel your hand get cold?
2. Why? (Alcohol evaporates, using body heat energy to change phase. This action is much quicker than for water, for alcohol has a low boiling point.)

Vocabulary: Evaporation, condensation, vapor

3.4 VARIABLES IN EVAPORATION

The volume, the initial temperature of water, and the size of the flame are constants. These are variables held steady so they will not affect the outcome. The surface area of the water is the variable. The shallow-wide dish allows water to expose more of its surface to the air. This allows for quicker evaporation as well as speedier absorption of heat energy from the same fires.

A key variable in evaporation is the humidity of air, its saturation with water vapor.

Vocabulary: Humidity, saturated, temperature

3.5 SOLUTIONS

This solution is a mixture of a liquid with another liquid. One can have mixtures of solids with liquids and gases in many combinations.

Answers:

1. By adding energy and heating the water.
2. Soap/detergent and water to wash dishes and clothes, sugar in coffee, tea, or hot chocolate, drink mix in water, frozen orange juice, soups, photographic chemicals, etc.

Mixtures: Mixtures are nothing more than a mixing of atoms, molecules, or compounds without permanent electrical bonds. Eventually, with some effort, these can be separated again. The properties of each member of the mixture are preserved. In a mixture of salt and water, the water can evaporate to leave behind the salt. Since all material objects are solid, liquid, or gaseous, one can have endless combinations of their mixtures.

Vocabulary: Solution, solvent, solute, soluble, insoluble

3.6 SPEEDY SOLUTIONS

You need a 35-plus cup coffee pot with hot water and a container full of ice water. You can add excitement to the activity by measuring the temperatures of waters. Provide the temperatures of waters to the class at the completion of all measurements. Notice that now students realize that they must record and graph everything. They also know that as an important element of any scientific investigation, they must plan their time wisely.

3.7 SCHLIEREN

Procedure: Place on top of an overhead projector a 300 ml beaker or a glass jar. Fill it with cold tap-water. Place one drop of a dark food coloring in the water. After a few seconds, line up all students and have them come by and see the molecular diffusion and the schlieren. Each student must take a good look at the beaker from its side, not the top. Ask students who are stiff to bend down a little! The overhead is needed to provide upward illumination of the phenomenon in progress, not for the projection on the screen.

Some vocabulary is omitted on purpose. It is introduced later with an appropriate experiment, to illustrate the meanings. Suggest that students observe what happens when they pour milk in tea or coffee, when at home. Another case of schlieren is visible when you travel on a highway on a hot day. You see the road ahead as if it were moving in waves.

Vocabulary: Fluid, turbulence

3.8 DISSOLVING SOLIDS

The solubility of solids in a liquid is enhanced by:

1. Increasing the temperature of the liquid
2. Increasing the quantity of the liquid

3. Increasing the agitation of the liquid

4. Increasing the surface area of the solute, for example, grinding solids into powder.

The limit to the solubility of any solid is the saturation point of the solvent.

Salt will reach saturation and temperature will have a negligible effect. Sugar will reach saturation and will very soon become the solvent, if the temperature is high.

Vocabulary: Solubility, residue

3.9 SUPERSATURATED SOLUTIONS: ROCK CANDY CRYSTALS

Stress that the jar for growing crystals must be extra clean. The supersaturated liquid, from which crystals grow, is the *mother liquor.* Crystal projects are rather difficult, because they take a long time. Have students check out books from the library on growing crystals.

Vocabulary: Precipitate, void, saturated, supersaturated

3.10 MOLECULES AND SOLUTIONS: OSMOSIS

The sugar or salt (NaCl, Sodium Chloride) first dissolves and forms a solution inside the bag. Next the solution passes out through small openings in the filter by osmotic action. The movement is from a region of higher pressure/molecular density to a region of lower pressure or density. Once on the plain water side, the salt solution dissolves even further. Gravity also helps the denser liquid to move downwards, though it moves out in all directions.

The visible action are the famous schlieren, lionized earlier! Now introduce the term *theory.* [A theory is an established explanation of many known facts or phenomena. Informally, it means a good guess, like a hypothesis.] In this investigation you see *circumstantial* evidence for the existence of molecules. You observe the results of molecular motion, not the molecules themselves. A good illustration of this are contrails in the sky. You know that jets and high flying airplanes make these, even though you cannot see the airplanes.

Vocabulary: Dense, membrane, microscopic, osmosis, theory, migrate, circumstantial

3.11 LIQUIDS: LAW OF CONSERVATION

You may need to extend on the law of conservation. Examples: $1.00 is still a dollar, even if changed into 20 nickels or 100 cents. If you divide a candy bar into several pieces, the candy bar is still there. Recall the definition of *system*: it is all there unless a part is taken away or added to it. If someone were to charge for the change, even if you only receive $0.95, the remaining coins remain in circulation (universe). This is preparation for when we must account for all energy during an energy exchange process.

Vocabulary: Conservation, liquid, molecule

3.12 GASES

Vocabulary: Invert, displacement

3.13 CARBON DIOXIDE GAS

If you have glass graduated cylinders of 50 ml, use them as the collection bottle. Students will be able to read the volume of carbon dioxide directly. Caution students to keep tablets dry. Place them inside plastic bags or wrap them in a piece of wax paper. To divide the tablets into quarters, score them with a razor blade or knife. Be patient, hand out extra tablets on demand. This is a wet investigation, and some students will make a total mess. Have several sponges and empty trash cans handy around the room.

Follow-up Activity: After students calculate gas for one tablet, let them do the investigation with one tablet, to verify their values. There is a space on the data model for this value. Make sure that the mouth of the bottle is large enough to handle one full tablet.

Vocabulary: Gas, generate, exhale, carbonated

3.14 IS WATER WET? ADHESION AND COHESION

Answers:

1. The water molecules hold together by cohesion and there is no adhesion force between the waxed paper and water.
2. This strong pulling together force is due to the adhesion between the glass and water.
3. Water wets glass and adheres to its sides, curving upward, forming a curve called the *meniscus.* Ice will eventually float to the edge, the highest point. Water does not wet plastic and bulges upwards (cohesion), and the ice cube floats to the center. Sometimes this activity will fail because some plastics conduct electricity.
6. The water wets the cup; the pulling force is greater than adhesion.

Vocabulary: Adhesion, cohesion, attract

3.15 WATER BUBBLE OR BUST! SURFACE TENSION

This is a neat experiment to show surface tension. It is easy to repeat and is a winner. Practice the penny demonstration and amaze your students.
In my class the scenario goes like this:

1. Fill your glass with water until not a single additional drop will fit, without overflowing. (This presents a challenge or dare).

2. When everyone has their glasses bulging at the top, incorporate the dropper activity. They will hardly believe that more is possible. You can add as many as 800 additional drops of water. Many of my students have done it consistently.

3. When the dropper has filled the glass to the maximum and one drop makes the difference, start your coin game. With practice you will be able to add about 50 pennies to the glass.

If you wish to burst the bubble, use a wetting agent like alcohol, or detergent, and no bubble will be possible. The wetting agent breaks the electrical bonds of cohesion, and the water becomes wetter.

3.16 COHESION: WETTING AGENTS

Caution: When working with surface tension activities, be extra careful to rinse out thoroughly (several times) all dishes and associated equipment. If any of the wetting agents from one activity remain behind, the next one will be a failure.

3.17 COHESION: SOAP BUBBLES

You may choose to purchase commercial bubble solution at a toy store for the final activity. To get a good spread in data, prepare small dropper bottles with a diluted soap solution (green tincture of soap USP). Use one part soap and three parts water in the dropper.

For giant bubbles mix liquid soap or detergent (Dawn™ works best), glycerin, and water. Glycerin is available at any drug store.

1. Mix one cup of detergent with ¾ cup glycerin and 8 cups of distilled water. You may wish to add more glycerin, water, or both. Experiment a bit. Make it into a class project.

2. Allow the bubble mixture to age overnight.

3. Prepare a large loop, 12 in or larger, but avoid wire coat hangers. These are coated to prevent rusting.

4. A cookie sheet or large tray works best as the container for the soap.

5. To make large bubbles, dip ring into soap solution. Make certain that entire ring is under the soap solution. Move forward rapidly to sweep ring and form large bubbles.

Caution: When the larger bubbles break, they leave a considerable amount of soap solution behind. Do this activity outside or over an easy-to-clean surface.

Information: Bubbles are really several bubbles in layers inside each-other. You have on the outside and the inside a bubble of water and in the middle a bubble of soap and glycerin. Wetting agents are available for reading glasses. These will help them to stay cleaner longer and to not fog. Most pharmacies have them. Get a bottle and demonstrate on a pair of student glasses. Cover one lens with anti-fogging (wetting) agent, then expose the glasses to the top of a steaming coffee pot. One lens will fog, the other will stay clear. Now coat the other lens.

A neat use of soap to break surface tension is to use it in your shower, bathroom, or wherever mirrors fog up. Cover these lightly with a damp cloth with a small amount of detergent or liquid soap added to it. When dry this coat will be invisible, but it will end the fogging.

3.18 THE VOLUME OF A DROP

You can use either a 10 ml graduated cylinder or a dropper with milliliters marked on it. These are available at pharmacies. Otherwise use a normal eye dropper with any small container marked in milliliters.

Answers:

1.–3. 1 ml divided by number of drops of alcohol = ml/drop of alcohol. Example: If 1 ml has 27 drops of alcohol, then $1/27 = 0.037037$ ml/drop of alcohol

4. $25 \times 0.037037 = 0.925925$ ml of alcohol
Suggest that students use calculators. You must specify how many decimal places need to be on the final answers. During the intermediate steps they should use all they have.

Name _____ Date _____ Period _____

3.1 THE MOTION OF MOLECULES

(Partner activity)

Purpose: The purpose of this activity is to observe the motion of molecules.

Information: All material objects, solids, liquids, and gases, are made up of tiny building blocks, which we call molecules. Molecules are so small that when you write a dot over the letter *i* with a pencil, you deposit on the paper millions of molecules of carbon. A Greek philosopher, Democritus (ca. 460–370 B.C.), tried to explain the nature of the world. He reasoned that material objects can be divided into half and then into half again and so on many more times. Finally one reaches the point when the material cannot be divided any more without losing its identity. This final indivisible chunk of matter is the *atom*. The atomic idea, developed by Democritus, was rediscovered in modern times.

Element is the word commonly used for two or more of the same atoms, since these rarely occur alone. Oxygen, gold, and copper are examples of elements.

Scientists use symbols to avoid writing out the names of elements. Examples: *Cl* for chlorine, *Na* for sodium, *C* for carbon. In writing symbols, scientists agree that the symbol represents just one atom, unless it has been changed by a number as a prefix or a subscript indicating more than one. Symbols sometimes do not agree with the English names of the element. The symbols come from the Latin or Greek names of the elements: *Na* (Natrium) for Sodium.

A *molecule,* the basic building block of matter, is the combination or union of two or more atoms. Examples: H_2O—water (composed of 2 hydrogens and one oxygen), NaCl—salt (composed of one sodium and one chlorine), O_2—Oxygen (composed of two oxygens). Molecules vibrate (are in motion) all the time.

Equipment: One bottom of a milk carton or plastic bottle nearly full with tap water, a metric ruler 30 cm (12 in), food coloring, paper towel

Procedure:

1. Prepare a data sheet like the model.
2. Set your dish on a table or better yet on the floor and fill it almost to the top with water.
3. Let the water stand 3 to 5 minutes or until it becomes still.
4. Carry the food coloring bottle inside a paper towel to your water dish.
5. Place one drop of food coloring in the water. Drop it from as close as possible to the surface of the water, but without touching it.
6. Observe and record carefully what the color does.
7. Measure and record the color drop at the start and then every minute for 5 minutes. Measure the drop at the bottom of the dish, at its widest point. Ignore the stain on the surface.

Questions:

1. Since the water was still and you did not stir it, what caused the dye to spread out?

_____

_____

2. Why is the water surface stain different from the stain on the bottom?

_____

_____

3. Why does the color blob appear to move to the bottom?

_____

_____

Sample Data Setup:

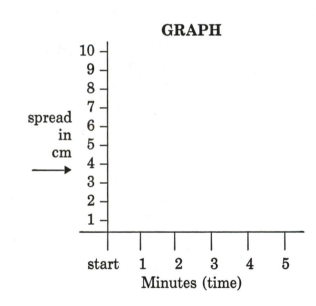

Time	Diameter of spread (cm)
start	
1 min.	
2 min.	
3 min.	
4 min.	
5 min.	

GRAPH

spread in cm →

10
9
8
7
6
5
4
3
2
1

start 1 2 3 4 5
Minutes (time)

3.2 PHASES OF MATTER

Purpose: The purpose of this investigation is to demonstrate the phases (states) of matter.

Information: All material objects are either solid, liquid, or gas. All matter is made up of molecules. These vibrate vigorously at a certain distance from each other, determined by their internal energy. Temperature determines the phase (state) of the material. Water is liquid at room temperature 72°F/22.2°C. On the North and South poles water freezes, forming large ice continents (a solid). At the equator water evaporates rapidly into water vapor (a gas).

To change materials from one phase (state) into another, heat energy must be either added or taken away. If molecules are close together, they form a solid mass. If they are farther apart, they form a liquid. If they are very far apart, then they form a gas.

Temperature is a measure of the average speed of molecules. The amount of heat energy present determines their speed. The more heat available, the faster the molecules vibrate and the more separated they become from each other. When you brew coffee, the evaporated water and coffee molecules become a gas that you can smell immediately. When you cut open an onion, some of its juices quickly evaporate into a gas you can smell throughout the house. Vibrating air molecules bump and mix with vibrating coffee and onion molecules, causing the mixture of gases to spread.

Equipment: Blow dryer, ice cube, dish, balance

Procedure:

1. Place an ice cube in a small dish and find its mass. Record it.
2. Make predictions on what will happen.
3. Write all predictions down.
4. Using a hair dryer, blow on the ice cube.
5. When the ice cube melts, reweigh the dish with the water.

Questions:

1. Has the dish with the ice cube changed its mass?

_____

2. If there is a change, why? Nobody has removed anything from the dish.

_____

3. What are the phases of H_2O?

_____

_____

4. Matter in the universe exists in three phases, each being dependent on temperature. Give examples of several other materials in different phases.

_____

_____

_____

_____

_____

3.3 EVAPORATION AND CONDENSATION

Purpose: The purpose of these activities is to demonstrate evaporation and condensation.

Information: To change water from a liquid to its gaseous phase, it must be heated. Anything above 0°C is enough; however, the higher the temperature the faster the process. Water boils at 100°C, and at this temperature the molecules change into gas the fastest. Going from a liquid to a gaseous phase is *evaporation.* When gases cool down and molecules change into a liquid phase, the process is called *condensation.*

Equipment: Balance, two cups or jars, hot and cold water

ACTIVITY 1: EVAPORATION: MOLECULES IN MOTION

Procedure:

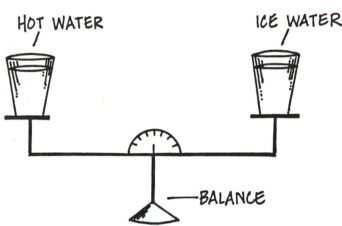

1. Place two similar dishes (glasses or cups) on the opposite pans of a balance.
2. Fill one dish with hot water and the other with ice water or cold water.
3. Balance the scale.
4. Predict what will happen and write it down.

ACTIVITY 2: COLD WATER EVAPORATION

Cold water evaporates too. You can see it by placing a dish with water on one side of the scale and balancing it with solids on the other pan. After a short while, the balance will tip down on the solid side.

Questions:

1. How are the two waters different?

_____

_____

2. How are the two waters different on the molecular level?

_____

_____

_____

ACTIVITY 3: EVAPORATION AND CONDENSATION

Equipment: Paper towel, water, glass, ice cubes, alcohol or acetone, aluminum foil

Part 1 Procedure:

1. Moisten a paper towel, wet the chalkboard and see the water disappear. (Make sure that the chalkboard is clean).

Questions:

1. Where did the water go?

_____

_____

2. Why?

_____

_____

3. What is this process called?

_____

_____

Part 2 Procedure:

1. Fill a glass with a couple of ice cubes and some water.
2. Let it stand and notice the frost on the outside of the glass.
3. Place several ice cubes in a container on the edge of a table and cover it with aluminum foil.

© 1991 by The Center for Applied Research in Education.

Questions:

1. Where did the moisture on the outside of the glass come from?

   _____

   _____

2. Where did the moisture on the aluminum foil come from? Explain.

   _____

   _____

3. What is condensation?

   _____

   _____

Homework: Place a glass in your freezer. After a few minutes take it out and observe the frost on it. Place some ice cubes in a glass with some water and observe the same condensation.

3.4 VARIABLES IN EVAPORATION

Purpose: The purpose of this investigation is to look at evaporation and its variables.

Information: If you wish to boil water more quickly, you use a wider pan in place of a regular narrow one. This allows a larger surface area of the pan to absorb heat energy, speeding up the process of heat energy transfer. The heat source is the same in both situations. In the last few investigations you saw water evaporate more quickly when it contained more heat energy. Other factors can influence the rate of evaporation.

Questions:

1. Name and write down everything that will influence the rate of evaporation of water. These are variables.

_____

_____

Equipment: One shallow and wide dish, one tall and narrow dish, balance, tap water, graduated cylinder

Procedure: Place the same volume of water in both containers and place these on the opposite pans of the balance. Balance the pans with any other object or masses.

Questions:

1. Predict what will happen with the balance.

_____

_____

2. Explain why you think this will happen.

_____

_____

Information: The National Weather Service has published this Apparent Temperature Table. It illustrates how heat combined with humidity creates an apparent temperature. The higher the relative humidity, the lower the rate of evaporation. This explains why on hot, humid days we feel hot and sticky. Our sweat, the body's air conditioning system, does not evaporate readily, and we feel very uncomfortable.

APPARENT TEMPERATURES
AIR TEMPERATURE (Degrees Fahrenheit)

Relative Humidity	70	75	80	85	90	95	100	105	110	115	120
	APPARENT TEMPERATURE (Degrees Fahrenheit)										
0%	64	69	73	78	83	87	91	95	99	103	107
10%	65	70	75	80	85	90	95	100	105	111	116
20%	66	72	77	82	87	93	99	105	112	120	130
30%	67	73	78	84	90	96	104	113	123	135	148
40%	68	74	79	86	93	101	110	123	137	151	
50%	69	75	81	88	96	107	120	135	150		
60%	70	76	82	90	100	114	132	149			
70%	70	77	85	93	106	124	144				
80%	71	78	86	97	113	136					
90%	71	79	88	102	122						
100%	72	80	91	108							

Questions:

1. What is the apparent temperature for 50% humidity and 90°F.?

_____

2. What is the apparent temperature for 70% humidity and 90°F.?

_____

3. What is the apparent temperature for 70% humidity and 95°F.?

_____

Name _____ Date _____ Period _____

3.5 SOLUTIONS

Purpose: The purpose of this activity is to introduce solutions and related vocabulary.

Information: When a liquid is placed into another liquid and it dissolves readily (creating a *solution*), it is *soluble*. Sometimes a liquid does not dissolve; it is *insoluble*. Sometimes liquids dissolve partially; they are *slightly soluble*. The liquid present in the largest amount is the *solvent*. The liquid present in the smaller amount is the *solute*. The *solution* that is the outcome of the dissolving is a *mixture*. Solubility or lack of solubility is a physical property of materials.

Equipment: One bottom of milk carton or plastic bottle half full with water, a metric ruler 30 cm (12 inches), food coloring

Procedure:

1. Prepare a data sheet as shown.
2. Take your water dish and place it on a table or better yet on the floor. The water will be still in about 3 to 5 minutes.
3. Place one drop of food coloring on top of the water. Be as close as possible to the surface, but do not touch it.
4. See what the food coloring does. Measure and record its overall spread for 4 minutes.
5. Repeat this investigation two more times and average the values.
6. Prepare a graph of these values.

Sample Data Setup:

Time	Diameter of spread (cm)		
	Try # 1	Try # 2	Try # 3
0 min.			
1 min.			
2 min.			
3 min.			
4 min.			

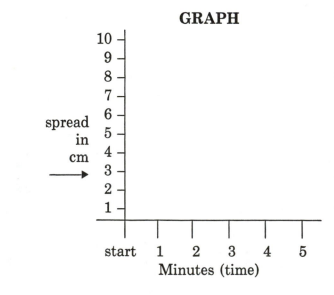

Questions:

1. Can you think of a way to speed up the rate of mixing in a solution, without stirring the liquids?

   _____

   _____

2. Can you name several mixtures that you have at home?

   _____

   _____

3.6 SPEEDY SOLUTIONS

(Partner activity)

Purpose: The purpose of this activity is to prove that heat energy makes molecules move faster.

Information: You can probably recall that molecules are in constant motion, or vibration. Molecules move with greater energy if you add energy to them. One simple way is to add heat energy.

Equipment: Three jars or milk cartons with equal amounts of water, about 6 cm to 10 cm of water (one container will have cold, one room temperature, and the last hot water), food coloring

Procedure:

1. Place one drop of food coloring into each jar of liquid.
2. Measure and record how far it spreads out by the end of 2 minutes.
3. Next, graph your data in bar graph form.

Sample Data Setup:

Temp.	Diameter of spread (cm)
cold	
room	
hot	

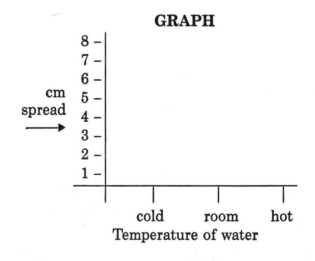

Questions:

1. Which food coloring drop spreads out the most? How much?

_____

_____

2. Which drop spreads out the least? How much?

_____

_____

3. Which spread had a middle value? How much?

_____

_____

4. Why did the largest one spread out the most? Explain.

_____

_____

3.7 SCHLIEREN

Purpose: The purpose of this activity is to create awareness of the word *schlieren.* This is another attribute.

Information: Schlieren are something which most people have observed, but they do not know what to call. Consequently they fail to describe certain situations in the appropriate manner. In science, schlieren are a common phenomenon. Schlieren are the visible streaks in a turbulent region in a fluid, with each streak (or schliere) having a density and index of refraction different from that of the greater part of the fluid. When you look at the bottom of a swimming pool, you observe the shimmering and changing lights dancing on the bottom. This is an outstanding example of schlieren. All those changes in the light are the direct result of light passing through different densities of the liquid.

Equipment: Glass jar, tap water, food coloring

Procedure:

1. Fill jar with tap water.
2. Place one drop of food coloring in jar.
3. Observe the coloring streaking down in streamers.

Equipment: Jar, water, food coloring

Procedure:

1. Students repeat experiment with glass jar, at their tables

Information: Schlieren photography is used to photograph turbulence in fluids (liquids and gases). You may recall seeing on TV smoke streams going past a car. These are schlieren produced by a wind tunnel.
   AT HOME: Pour some cold milk into hot tea (in a clear glass) and observe.

Questions:

1. Explain in your own words: What are schlieren?

   _____

   _____

2. Provide a few examples of schlieren.

   _____

   _____

Name _____ Date _____ Period _____

3.8 DISSOLVING SOLIDS

(Partner activity)

Purpose: The purpose of this activity is to become well-informed about the process of solution.

Information: If a solid dissolves in a liquid, the solid is *soluble*; otherwise, it is *insoluble*. Baking soda, salt, and sugar are soluble in water. Glass, wood, and metal are insoluble in water. The substance which is present in the greater quantity in a solution is the *solvent*. The lesser material in a solution is the *solute*. Some solids are only partly soluble.

Equipment: Two test tubes with support rack or baby food jars or plastic glasses, salt, sugar, room-temperature, cold, and hot water, teaspoon, ruler, waxed paper (to make funnel to pour solids into test tube).

Information: You will test the solubility of two different solids in a liquid. While you perform three separate experiments, you will use the same amounts of water at the same temperature. You will stir the containers equally, to keep the variables (temperature and stirring) the same. If under the same conditions one solid leaves less residue on the bottom, that is the evidence of its greater solubility.

Note: If you use test tubes, fill them half full with water. Use two level teaspoons of sugar or salt. Sometimes you will need a few more. Make a small funnel with the wax paper to pour the salt/sugar into the test tubes.

If you use baby food jars or similar clear containers, fill half full with water. Add 5 to 15 teaspoons of salt and later repeat with sugar. Test a few times until you get definite residues on the bottom.

Procedure:

1. Place salt in a container half full of room temperature water.
2. Shake or stir the container 15 times and let stand for a couple of minutes. All undissolved solids will settle to the bottom.
3. Measure the height of the residue. In wider containers, estimate residue by teaspoon.
4. Record all data.
5. Repeat using cold water.
6. Repeat using hot water.
7. Redo steps 1–6, using sugar.

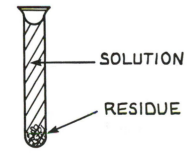

Sample Data Setup:

Temp. of H_2O	Undissolved	
	sugar (mm)	salt (mm)
cold		
room		
hot		

GRAPH

undissolved material (mm) solution residue

8 –
7 –
6 –
5 –
4 –
3 –
2 –
1 –

cold   room   hot
Temperature of water

Questions:

1. Which solid dissolved more readily?

_____

_____

2. What is your evidence? Quote specific measured values for evidence. Never ask the reader to look up data. You are making key scientific points by using important supporting evidence you have developed.

_____

_____

3. What factors improve the solubility of a solid in liquids?

_____

_____

4. Is there any other way to improve the solubility of solids in liquids?

_____

_____

5. Can you design another lab to check your answers in question 4? Explain.

_____

_____

3.9 SUPERSATURATED SOLUTIONS: ROCK CANDY CRYSTALS

Purpose: The purpose of this activity is to introduce crystal growth.

Information: In the last investigation, you discovered that the solubility of substances improves with stirring (adding mechanical energy) and by heating (adding heat energy). You have found that the solvent (water) dissolved just so much solute (salt or sugar). After a time, the excess solute precipitates to the bottom. When this happens, the liquid is a saturated solution.

Materials dissolve in larger quantities when placed into heated liquids because the molecules of these liquids vibrate further apart. The spaces between the molecules become greater and more solid molecules (sugar and salt) can fill these voids. As the liquid cools down, the spaces between molecules become smaller and the excess solids precipitate to the bottom. Heating solvents to dissolve additional materials allows one to obtain a supersaturated solution. We use supersaturated solutions to make gelatin and many crystals used for computers and the electronics industry.

Molecules are always vibrating, unless one removes their energy by cooling them to absolute zero. To date, scientists have been unable to reach absolute zero, which is $-273°C$, or $-460°F$.

Equipment: Cooking pot, measuring cup, sugar, water, clean empty quart jar, sewing thread, pencil or paper clips

Procedure:

1. Put two cups of water into a clean pot. Use a measuring cup.
2. Bring the water to boil.
3. While the water is heating, add tablespoon after tablespoon of sugar. Be careful not to add any sugar unless the previous tablespoon has dissolved. Use 5.5 cups of sugar.
4. Stir to help dissolve the sugar. When no more sugar dissolves and the liquid mixture is thick and boiling, the solution is supersaturated.
5. Turn off the heat and let the pot cool until it is barely lukewarm, almost cool to the touch.
6. Pour the supersaturated solution into the clean jar.
7. Do this step only after the liquid in the jar has completely cooled down. Suspend a Life Saver or other hard candy from a string, about 3 cm from the bottom.
8. Use a pencil or open a paper clip to support the string on top of the jar. Twist the pencil, to adjust the height of the candy.
9. Cover jar with a piece of waxed paper to keep dirt out. Avoid jarring the liquid. As the liquid sits, the excess sugar molecules will come out of solution and will remain on the string and the seed candy. After a few weeks, bring your rock candy to school. Watch out for ants! Adding food coloring makes it harder to see the growing crystals.

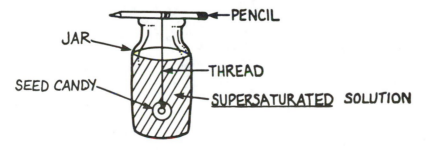

3.10 MOLECULES AND SOLUTIONS: OSMOSIS

(Partner activity)

Purpose: The purpose of this activity is to demonstrate the principle of osmosis.

Information: So far you have seen molecules in slow motion. Now is the time to see some fast molecular action. When you have two different solutions, the molecules from the denser one migrate toward the less dense one to form a mixture. If the two solutions are separated by a membrane, this movement takes place through the microscopic holes in the membrane—this is called *osmosis.* When you receive a shot in the arm, the medicine travels through your body to where you need it. Your body acts as a membrane.

Equipment: Clear and clean jar nearly full with fresh tap water, a teaspoon of sugar or salt, one coffee filter

Procedure:

1. Place sugar or salt into the filter paper and squeeze it to form a bag.
2. Wet the top of this bag.
3. Place it into the jar with its upper part folded over the jar lip, for support.
4. Observe the action, from the sides. Record everything. With your instructor's permission, taste the water after the action stops.

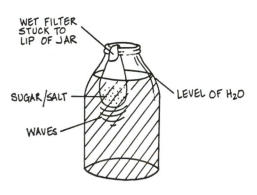

Questions:

1. Does water go into the filter bag? Is there any evidence of this?

_____

_____

2. What happens inside? What is your evidence of this happening?

_____

_____

3. Did anything happen outside the bag? Describe everything.

_____

_____

4. Did the sugar/salt move outside the bag? Your evidence?

_____

_____

5. What is this process called?

_____

_____

6. Did the sugar/salt dissolve? Did you notice waves or stringers? What causes them?

_____

_____

7. Did anything come out of the bag?

_____

_____

8. What causes the sugar/salt to dissolve?

_____

_____

3.11 LIQUIDS: LAW OF CONSERVATION

(Partner activity)

Purpose: The purpose of this activity is to notice the liquid phase (state) of matter, while learning about the law of conservation of matter.

Information: Matter comes in four phases (states): solid, liquid, gas, and plasma. All matter is made up of molecules. Liquids have the following properties:

1. They take the shape of their container.
2. The molecules of liquids slide around each other.
3. The amount of the liquid remains the same, even if you pour it from one container into another.

Equipment: Several bottles and jars with different shapes, jar, water

Procedure:

1. Fill one jar with water to the brim and look upon it as the master container.
2. Pour this water into three or four other containers.
3. Next pour all the water back into the master container.
4. Repeat this activity by using a variety of bottles as master containers.

Questions:

1. Were you able to refill the master container, when you returned the liquid back?

_____

_____

2. Did you get the same results for different master containers?

_____

_____

Information: Since the divided liquids filled the original container, this is conservation. The law of conservation states that matter or energy cannot be either created or destroyed. However, it does allow that matter and energy may change form. Example: An ice cube melts. It turns from solid to liquid and maybe into water vapor, but all the molecules are still in the universe.

3.12 GASES

(Partner activity)

Purpose: The purpose of this activity is to learn properties of gases while learning to use the water displacement method.

Information: Gases take the shape of the containers they are in. One cannot see most gases, for they are colorless and sometimes tasteless and odorless. Examples: air, oxygen, helium, carbon monoxide, and natural gas.

Equipment: One dish pan or fish tank, two clear plastic glasses

Procedure 1:

1. Fill dish pan or fish tank with water, not too full.
2. Place inverted glass into water.

Questions:

1. Describe what you see.

_____

_____

2. What will happen if you invert the glass under the water?

_____

_____

Procedure 2:   Invert the glass under the water.

Questions:

1. What did you see?

_____

_____

2. Is your observation evidence that there is gas trapped in the glass?

_____

_____

3. What kind of gas is trapped in the glass?

_____

_____

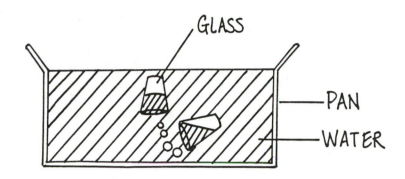

GLASS

PAN

WATER

Procedure 3:

1. Place one glass under the water. Fill it with water.
2. Hold the other glass upside down under the water. The trapped air remains under the glass.
3. Lift the glass full with water and hold its bottom below the surface of water. (See above.)
4. Tilt the glass with air and let some of the air escape. Let the escaping air enter the glass with water. Keep the glass that was full with water pointed down and above the escaping air.
5. Let all air escape until it fills the other glass.
   This process is the *water displacement method.* Scientists use this method to collect gases during experiments.

Name _____ Date _____ Period _____

3.13 CARBON DIOXIDE GAS

(Partner activity)

Purpose: The purpose of this activity is to learn more about gases.

Information: Humans and animals exhale many gases, including carbon dioxide and water vapor. To see the water vapor, exhale quickly on a mirror in front of your mouth. You will see the moisture fogging it. Carbon dioxide is also released when fossil fuels and wood burn. In theory this might produce the greenhouse effect: heat is received from the sun but it is trapped on our planet by the carbon dioxide in the atmosphere. Heat cannot be radiated back into space, so the temperature on earth increases, just as in a greenhouse.

Carbon dioxide is an invisible gas. Its chemical symbol is CO_2. When vinegar and baking soda react, the bubbles of the released gas are CO_2. Some other CO_2 sources are: the fizz in carbonated drinks; Alka Seltzer™, combining with water; and dry ice, which is CO_2 in solid form.

Equipment: One small beverage bottle, one quart bottle, dish pan and five (ea.) ¼ tablets of Alka Seltzer™, 100 ml graduated cylinder, waxed paper, masking tape.

Procedure 1: This is a trial run.

1. Fill a small pie tin or pan with water about 3 cm to 4 cm deep.

2. Place a piece of masking tape from the top to the bottom along the entire height of the bottle.

3. Fill the bottle with water, cover it with a piece of waxed paper, and invert it.

4. Place the bottle's neck under water in the pan. Remove the wax paper.

5. Quickly place a piece of Alka Seltzer™ tablet under water into the neck of bottle.

6. Collect the reaction gases. This is collection of gas by the water displacement method.

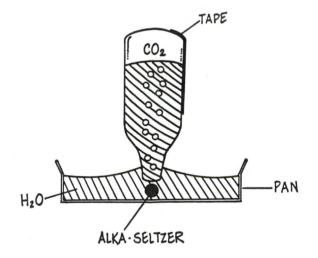

Procedure 2:

1. Using the water displacement method, (steps 1–5) collect the gases produced by ¼ tablet of Alka Seltzer™.
2. Mark the height of the gases on the outside of bottle.
3. Empty the bottle and then fill it up with water to the mark you just made.
4. Carefully pour this water into a graduated cylinder and measure its volume. You measure the volume of water equivalent to the gas generated by ¼ tablet.
5. Record this volume and repeat the investigation two more times.
6. Average the answers to minimize errors in your measurements.

Sample Data Setup:

fraction of tablet	volume (ml) try 1	volume (ml) try 2	volume (ml) try 3	average (ml)
¼				
1 tablet		(optional)		

Questions: The data you get from this investigation will help you find out the average volume of gas produced by ¼ tablet of Alka Seltzer™.

1. How much gas can be produced by a whole tablet? Show your calculations.

_____

_____

2. How much gas can be produced by 25 tablets? Show your calculations.

_____

_____

3.14 IS WATER WET? ADHESION AND COHESION

(Partner activity)

Purpose: The purpose of this activity is to demonstrate adhesion and cohesion.

Information: Everyone believes that water is wet because it sticks to material objects like paper, wood, metal, and skin. The property of water sticking to other substances is *adhesion*. *Adhesion* means that the electrical charges of *one* material attract and are attracted by those in *other* materials. Wetting depends on adhesion. When adhesion forces are low, the electrical charges of water do not attract those of other materials and water does not wet. Wetness is defined in science as the angle between the substance and the wetting material. Examples of non-wetting are water coming in contact with fats, greases, oils, and most plastics.

Cohesion occurs when molecules in *one* material do not separate, but stick together because of their electrical attraction for each other. Example: siphoning water or gas. The water or gas comes out of a hose in a stream, and the cohesion of water/gas molecules pulls the liquid out of the container. Think of cohesion as molecules in a material resembling the links of a chain.

Equipment: A piece of waxed paper, water, two microscope slides, one plastic and one glass cup, two ice cubes, empty aluminum can, flat pan.

Procedure 1: Sprinkle some water on the waxed paper.

Question:

1. Why does the water not spread out all over the waxed paper?

_____

_____

Procedure 2:

1. Place a drop of water between two microscope slides or two flat pieces of glass.
2. Press the slides together and slide them apart so they have a central overlap area.
3. Try to lift them away from each other without sliding. You will notice that they resist pulling apart.

Question:

2. Why are the pieces of glass hard to separate? Explain.

_____

_____

Procedure 3:

1. Fill both cups (one plastic, one glass) nearly full with tap water and place an ice cube in each.
2. Wait a few minutes.
3. Record what you see.

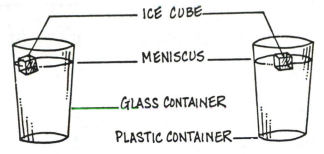

Question:

3. Why does the ice cube stay on the edge of one cup and in the center of the other?

_____

_____

Procedure 4:   Take any standard paper cup which has a small ridge going beyond the bottom of the cup and place it into any container with water. Pull it out gently and feel the resistance.

Questions:

4. Were you able to pull the cup out of water?

_____

5. Did you feel any pull against you?

_____

6. Why were you able to pull the cup out of the water?

_____

_____

_____

3.15 WATER BUBBLE OR BUST! SURFACE TENSION

Purpose: The purpose of this activity is to continue demonstrating cohesion.

Information: Water sticks to the sides of a glass container due to adhesion. Minute electrical charges in water molecules attract those in glass. You must have noticed that small objects like paper, dust, and insects float on water as if it had a surface skin. The molecules on the surface of water electrically attract each other with much energy and produce a skin: surface tension. Cohesion is the electrical force that causes this bonding.

The cohesion of water molecules allows you to fill a container with water more than normal, since surface tension holds water in. When the inner molecular pressure of water exceeds the holding force of surface tension, the bubble (surface tension) bursts.

Equipment: Flat aluminum pan, glass or jar, about 50 pennies, water, eye dropper, water

Procedure:

1. Fill glass or jar with water to the very top, until no more can be added. Notice the fairly high bubble above the glass, looking at it from the side.

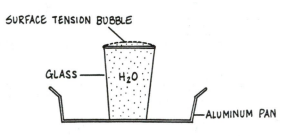

2. Use an eye dropper to add more water to the glass, counting each drop.

3. Record how many drops of extra water you can add before the bubble bursts.

4. When the glass does not take any more water, drop pennies, one at a time, into the glass from the side. Hold pennies with their faces vertical. You will be surprised at how many pennies will go into the glass without the bubble bursting. Drop pennies very close to the surface of water. Do not touch the surface of the water with your fingers while holding a penny.

5. Record the number of pennies you dropped into the container.

Sample Data Setup:

TRY #	CONTAINER	EXTRA DROPS OF WATER	EXTRA PENNIES
1			
2			
3			

AVERAGE ⟶ _____   _____ EA.

Questions:

1. What was the average number of extra drops of water that you were able to add to the glass?

_____

2. How many extra pennies were added after you filled the glass with a dropper?_____

3. Explain what is meant by the following terms: a. adhesion b. cohesion c. surface tension

_____

_____

_____

_____

_____

_____

3.16 COHESION: WETTING AGENTS

(Partner activity)

Purpose: The purpose of this activity is to observe the property of cohesion in action.

Information 1: Most scientists consider water as the universal solvent. Water owes its surface tension to cohesion. Cohesion affects anyone who uses water. Water refuses to adhere (stick) to fats, plastics, skin, plates, laundry etc.

By adding soap to water, you partly neutralize the electrical bonds in water. Adhesion increases and water becomes wetter than water without soap. Now it can really do the cleaning jobs, since it spreads out evenly over greasy and plastic surfaces. Wetter water is critical in photography so films and pictures do not show water stains from their processing. Firefighters add wetting agents to water to make it flow faster through their hoses. Soap, detergents, rinses, and alcohol are wetting agents because they condition liquids to be wetter. Wetting agents are used in medicines, cosmetics, and many chemical products. Salad dressings use wetting agents to improve the mixing of oil and vinegar. Soap and detergents surround fat particles in water. This is *emulsification.* You will recognize emulsification because fats mixed with water turn the liquid milky white.

Equipment: Several pins, pepper or talcum powder, paper clips, tweezers, string, shallow pan with water, dropper, dropper bottle with soap or alcohol

Procedure 1:

1. Fill your pan about half full with water and take it to your seat.
2. Using the tweezers, place gently on the water surface several pins and one paper clip. Make sure that they float on the water surface.
3. Place one drop of wetting agent on the water surface.

Questions:

1. Why do objects float on water?

   _____

   _____

2. Why does your hair stick together after you wet it during a shower?

   _____

   _____

Procedure 2:

1. Rinse out your dish several times, refill it, and return with it to your seat.
2. When the water has become still, the teacher will sprinkle the surface of water with pepper or talcum powder.
3. Place one drop of wetting agent on the surface of the water.
4. Observe and record.

Procedure 3:

1. Tie several loops of string. Use string about 10 cm long.
2. Rinse your dish out carefully, refill it, and return with it to your seat.
3. Place a loop on top of the water. Drop one drop of wetting agent in the middle of the loop.
4. Observe and record.
5. Repeat with a different loop.
6. Rinse out your equipment and refill it.

Procedure 4:

1. Take a piece of paper and roughly fold it into the shape of a boat. Make it about 1 cm to 2 cm long.
2. Cut a V-notch in the back of this boat.
3. Float boat in water.
4. Add drop of soap in the V-notch. Watch the boat sail in the dish.

Procedure 5:

1. Take two toothpicks and place them together in the water.
2. Place one drop of soap between the toothpicks.
3. Record all observations.
4. Wash all equipment and repeat this last experiment, using a drop of alcohol instead of soap.

Information 2:   You need to find out which has the greater volume: a drop of water or a drop of alcohol. You will repeat this investigation several times and average the values to reduce errors in measurement.

Procedure 6:

1. Use a dropper to count how many drops of water fill a plastic spoon to the brim. Hold the dropper vertically above the spoon.
2. Repeat two more times and average the values. When averaging, round off to the nearest whole number.
3. Repeat steps 1 and 2 with alcohol.

Sample Data Setup

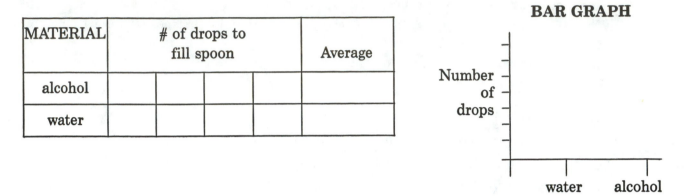

MATERIAL	# of drops to fill spoon				Average
alcohol					
water					

BAR GRAPH

Number
of
drops

water   alcohol

Questions:

1. Which drop has the greater volume, water or alcohol? Explain why.

_____

_____

2. Why do you repeat an experiment several times and average the results?

_____

_____

Name _____ Date _____ Period _____

3.17 COHESION: SOAP BUBBLES

(Partner activity)

Purpose: The purpose of this activity is to examine cohesion of molecules, using soap bubbles.

Information: The cohesion bond between water molecules increases with the addition of soap or detergents. If you place a drinking straw in water which contains no soap, you cannot blow bubbles. If you add soap to the water, bubbles are possible.

Equipment: Green tincture of soap or liquid detergent in small dropper bottle, small cup, drinking straw, distilled or demineralized water, graduated cylinder

Procedure:

1. Place one drop of soap or detergent in a small cup with 50 ml of distilled water.
2. Place straw in liquid, stir gently, and pull it out.
3. Gently blow to form bubble.
4. Add another drop of soap until you can blow the largest possible bubble.
5. Every time you add a drop of soap, measure the bubble and record it. Blow a few bubbles and record the largest.
6. Compare your largest bubble with one you will make with a soap solution provided by the teacher.

Sample Data Setup:

number of soap drops	size of bubble (cm)
(Example) 1	0 cm

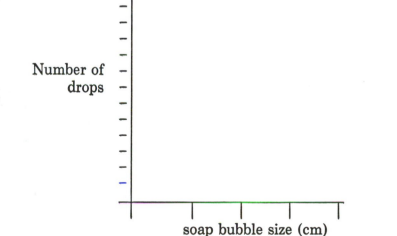

GRAPH

Number of drops

soap bubble size (cm)

Questions:

1. How many drops of soap do you need to make a large soap bubble?

   _____

   _____

2. How does this bubble compare with the one you make using the soap solution provided by the teacher?

   _____

   _____

3. Why do you think bubbles do not form when there is no soap in the water?

   _____

   _____

3.18 THE VOLUME OF A DROP

(Partner activity)

Purpose: The purpose of this activity is to measure and calculate the volume of drops of different liquids.

Equipment: Specially marked eye dropper or 10 ml graduated cylinder, a flat dish, one eye dropper, alcohol, vinegar, water.

Procedure:

1. Count the drops of alcohol needed to fill the smallest unit of your measuring device. The goal is 1 ml. If you have a specially marked eye dropper in ml, count how many drops of alcohol there are in one ml.
2. Repeat several times to obtain greater accuracy. Average to nearest whole number.
3. Divide 1 ml by the average number of drops in a milliliter. You get the volume of a single drop of alcohol.
4. Repeat steps 1 through 3 for water and for vinegar.

Sample Data Setup:

MATERIAL	number of drops in 1 (ml) or cc				average # of drops per ml
ALCOHOL					
WATER					
VINEGAR					

Questions: NOTE: Answer all questions and include a separate sheet to show all your calculations.

1. What is the volume of one drop of alcohol?

_____

2. What is the volume of one drop of water?

_____

3. What is the volume of one drop of vinegar?

_____

4. What is the volume of 25 drops of alcohol?

_____

5. What is the volume of 20 drops of water?

_____

6. What is the volume of 15 drops of vinegar?

_____

4

CHEMISTRY

TEACHER'S SECTION

In this section students will learn and perform activities on:

1. Atoms and elements
2. Structure of atoms
3. Mixtures
4. Compounds and molecules
5. Chemical and physical changes
6. Evidence for oxygen
7. Percent of oxygen in air
8. Acids/bases and litmus paper
9. Litmus paper and hydrion
10. Bromothymol blue and bad breath
11. Testing for starches with iodine
12. The hardness of water
13. Unknown powders: variables
14. Nuclear energy: fission
15. Splitting an atom

Safety Note: When using candles, avoid personal liability by doing the activities yourself and by having students measure your work. Some states forbid candles in elementary classrooms. When handling chemicals, use and provide aprons, goggles, and gloves when appropriate. Follow all local, state, and federal safety regulations.

4.1 ELEMENTS

Vocabulary: Symbol, process

4.2 ATOMIC STRUCTURE

Vocabulary: Subatomic, proton, neutron, electron

4.3 MIXTURES

Water in its pure state is a compound, H_2O, but regular tap water is a mixture. It contains many minerals: Fe (iron), Ca (calcium), Mg (magnesium), Cl_2 (chlorine), and probably many other chemicals. When metals mix, alloys result. Gold, due to its softness, is alloyed with either copper or platinum. The percentage of noble metals (gold, platinum, etc.) in alloys is expressed in karats. Pure noble metal is designated 24 karat, while 18 karat means 75 percent.

Air is a combination mixture and compound, for it contains H_2O—a compound of hydrogen (H) with oxygen (O_2)—Nitrogen (N_2), carbon monoxide (CO), H_2O—a compound of hydrogen (H) with oxygen (O_2)—nitrogen (N_2), carbon monoxide (CO), carbon dioxide (CO_2), neon (Ne), and Argon (Ar).

Demonstration Activities:

1. Dissolve some salt in a beaker (a mixture), then evaporate the water.
2. Mix some beans and peas. Have students separate them.

Vocabulary: Mixture

4.4 COMPOUNDS AND MOLECULES

Answers:

1. 5
2. Iodine
3. Oxygen
4. Sugar 35
5. They all have oxygen and are oxides.

Vocabulary: Oxide, oxidation, metabolism, rust, fire, explosion, endothermic, exothermic

4.5 CHEMICAL AND PHYSICAL CHANGES

Demonstration: Preparing a Pancake

1. Prepare a pancake batter. It is a mixture until you cook it. If you add leavening to it, you start a chemical reaction.
2. Cook the batter. You have a new substance, which will not change back into the liquid batter. This is a chemical change. You needed heat. This is an endothermic chemical reaction, typical of most foods.

An *Endothermic* reaction takes in heat energy as it occurs. An *exothermic* reaction gives off heat energy—examples of this include sink trap cleaners and candles burning.

Answers to activities:

a. physical change	g. physical change	m. chemical change
b. physical change	h. physical change	n. physical change
c. physical change	i. chemical change	o. physical change
d. physical change	j. physical change	p. physical change
e. chemical change	k. physical change	q. chemical change
f. chemical change	l. chemical change	

Answers to questions:

1. chemical change
2. physical change
3. chemical change
4. chemical change

5. physical change
6. chemical change
7. physical change
8. physical change

9. physical change
10. chemical change

Vocabulary: Chemical change, physical change

4.6 EVIDENCE FOR OXYGEN

Oil is a barrier to keep oxygen from rusting (oxidizing) the steel wool. Paint is a barrier to keep oxygen from oxidizing metals. Note that sometimes metals are covered in thin layers of non-oxidizing metals like gold, platinum, chrome, etc.

4.7 PERCENTAGE OF OXYGEN IN AIR

Oxygen in a combined state (forming molecules) makes up one half (50%) of the earth's crust and two thirds (66%) of the mass of human and other animal bodies. Oxygen in its free state makes up about one fifth (21%) of the volume of air. Fish get oxygen by using the air dissolved in water (about 10% of the water). Joseph Priestly, an English chemist (1733-1804), discovered oxygen. Antoine Lavoisier (1743-1794), a French chemist, named it. He was the first to explain oxygen's role in burning, breathing and oxidation of metals. Oxygen is a colorless, odorless, tasteless gas, slightly heavier than air (1.07%), slightly soluble in water and convertible into liquid form (LOX). All gases can be *liquified* by cooling and compressing. *Oxidation* takes place at many speeds:

- slow (rusting)
- fast (burning)
- extra fast, instantaneous (explosion)

During oxidation heat is given off. If the heat cannot be given off, it is accumulated until a sudden expansion results in an explosion.

Plants use carbon dioxide from air and return oxygen. Oxygen is used to make a hot flame in welding. In emergencies, people are given oxygen to help them breathe.

Note: To get good data, try to use jars which are cylindrical and do not have a narrowing at the neck. You need an even column of air.

Calculations: Provide class with calculators. The formula you use is the typical percentage formula: Part/Whole $\times$ 100 = Percent.

Answers:

1. The bubbles escape because the candle heats the air in the jar and the air expands.

2. The water rises because the oxygen burns. Inside the jar there is less air pressure because the oxygen has been burnt. The weight of the atmosphere pushes the water inside the jar, until the inside and outside pressures are equal.

4.8 ACIDS, BASES, AND LITMUS PAPER

Get litmus paper and pH pencils from any science supply house. One pencil set is good for thousands of tests and many take-home activities. Any surface becomes a test indicator—just make a mark. Moisten the surface, and in less than 15 seconds the indicator turns color. Follow this activity with five or six unknown liquids in class: water, ammonia, alcohol, detergent, vinegar, and so forth.

Point out that red litmus paper is actually pale pink, and blue litmus paper is pale violet.

Answers: Acids turn all litmus red. Bases turn all litmus blue. Neutral substances change neither, so you must test with both colors. Obviously, if a litmus strip is already the right color, nothing changes.

LITMUS PAPER TESTS

	RED	BLUE
ACID	RED	RED
BASE	BLUE	BLUE

A sample rhyme:

Acids show red
And bases blue
Whenever a litmus strip
Changes its hue.

Vocabulary: Acid, base, buffering, indicator (chemical), organic, caustic

4.9 LITMUS PAPER AND HYDRION

Note: Define *organic* as a substance containing the element carbon. This is evidence that it comes from a living material, such as a plant.

Procedure 1:   Set up several small dishes with clear or nearly clear liquids, for example, water, diluted vinegar, diluted ammonia, soda water, water with baking soda, and water with detergent. Place each small dish in a flat, shallow pie tin. Label each sample by number only. Do not disclose the contents until after the activity.

Procedure 2:   A neat follow-up is to treat a standard notebook page with ten to twelve lines of hydrion pencil.

1. Ask students to line up single file.

2. It will take you seconds to make ten to twelve marks on each paper with the pH pencil.

3. Students can go home and test several additional substances to find out whether they are acid, alkaline, or neutral.

4. Ask students to write below each indicator line the substance tested and the result. They must take a reading within 10 to 20 seconds of starting. The paper underlying the hydrion strip is acid. Given enough time this will affect the result.

Giving out litmus strips creates problems. Students eat them, lose them, wet them, etc. If you decide to use litmus strips, send them home in plastic bags. Be comfortable in knowing that the hydrion pencil is nothing more than bromothymol blue in pencil form, in its acid mode, which makes it appear yellow. Some brands of this pencil appear green, for bromothymol is in its neutral mode. It may change color between your application and the test time at home. The results of testing will still be credible.

This is a particularly enjoyable project for most students. Follow up with a survey of results. Note that shampoos can be acid, neutral, or basic. Water is slightly basic, about pH 7.4. This may vary for different regions of the country. Note that most foods are acid. This result is not an error due to the indicator or other variables.

Vocabulary: Acid, base, neutral

4.10 BROMOTHYMOL BLUE VS. BAD BREATH

Bromothymol blue changes into light green and then into clear yellow in the presence of acids. If you add a base, it will return to its blue condition. Test with vinegar and baking soda. In this investigation, we verify that one's breath is acid. This acid is partly to blame for bad breath, and no amount of mouth washing can mask it effectively. Additional causes of bad breath: eating garlic or onion, tooth decay, poor health, poor oral hygiene.

4.11 TESTING FOR STARCHES WITH IODINE

Use aprons. You can use all kinds of familiar materials for testing: bread, potatoes, apples, crackers, canned foods, chalk, fruit juices, rice, marshmallows, popcorn, baking soda, an ordinary sheet of paper, paper towels, the paper plate used, and so forth.

Caution: If you use a tincture of iodine, USE WITH CAUTION—IT IS TOXIC. Water-soluble iodine, the type commonly used by hospitals (Betadine™ or Wescodyne™) is nontoxic. If your students get iodine stains on their clothes despite aprons, use photographic fixer (sodium hyposulphite) to remove them.

Vocabulary: Thickener, filler, stiffener, glucose

4.12 WATER HARDNESS

Prepare three or more bottles, each clearly labeled, with tap water, soft water, and sea water. If you do not live by the sea, add as much salt as you can to the water in the last bottle to get a saturated solution. Add also a couple of tablespoons of baking soda to simulate minerals. Use tincture of green soap USP available from any pharmacy. Dilute about 10:1 and use in plastic dropper bottles. Water hardness is rated in grains of hardness. One grain equals 17 parts per million (ppm) of calcium carbonate. Soft water has 0 grains of hardness. Some of the hardest drinking waters have over 340 ppm.

Vocabulary: Mineral

4.13 UNKNOWN POWDERS: VARIABLES

Fill jars with the following (in any order you wish) and label with a number (label the lids and bottoms of jars, for later match).

1. granulated sugar
2. baking soda
3. cornstarch
4. talcum powder
5. baking powder
6. chalk dust (or obtain white marking powder from P.E. dept. or calcium carbonate ($CaCO_3$) in pharmacy)
7. white (wheat) flour
8. salt
9. confectioner's sugar
10. baking powder
11. powdered antacid tablet
12. your choice

Make certain that students have completed the first part (steps 1 through 6) before starting on the final part. In the first part, students will learn about specific powders and identify their properties. Extending the activity to additional unknowns will provide closure to this activity. At the end, reveal the jar contents to students.

Vocabulary: Reagent

4.14 NUCLEAR ENERGY: FISSION

This activity is risky with younger children, due to the unstable nature of mousetraps. You may wish to videotape this setup for future use. For a 25 square foot area you need 515 mousetraps, which at $.33 each come to about $170, plus the ping pong balls. Maybe you can borrow these from a hardware store or join the local ping pong club. I assure you that this demo is impressive. Before videotaping, place mirrors on the corner walls to reflect and enhance the video. Shoot from a low angle, near the ground.

Vocabulary: Nuclear, radioactivity, fission, repel

4.15 AN OIL-DROP MODEL OF SPLITTING AN ATOM

This activity lends itself to a grand teacher demonstration. Use a videocamera, if available, and show on several screens. A good follow-up is assigning it as homework to practice. Have students show in class their mastery of this demonstration.

Nuclear Fission: When a neutron enters the nucleus of an atom, it splits the nucleus and releases the binding energy of the nucleus together with heat and all subatomic particles. Consider that the nucleus has protons, positive charges which repel each other. Ergo the nuclear force that binds it must be extremely large.

Nuclear Fusion: At or about 10,000,000° F atoms literally fuse together. When four hydrogens fuse together, helium is the product. You will notice that the sum of four hydrogen masses is slightly larger than the mass of a helium. The difference is the amount of energy released. Note that this is another beautiful example of the law of conservation of mass and energy.

The sun is a prime example of thermonuclear fusion in process.

Answers: Fission and fusion.

Vocabulary: Deformation, fusion, fission

4.1 ELEMENTS

Purpose: The purpose of this activity is to introduce students to elements and the periodic chart.

Information: For a listing of all elements, you need to look at a Periodic Table of Elements. In chemistry, we use symbols to abbreviate the names of elements. The symbols represent an element and a definite mass of that element. Example: *H* is the symbol for hydrogen, and its atomic mass is 1; *C* is for carbon, and its atomic mass is 8; *O* is for oxygen, and its mass is 16. To avoid decimals, atomic masses are rounded off to the nearest whole number. Note that if the atomic mass of H (hydrogen) is used as the standard, one atom of C (carbon) has a mass eight times greater than one H.

The symbol for an element is made up of one or two letters. The first letter is always capitalized. Two letters are used to avoid confusion when two or more elements begin with the same letter, such as *S* (sulfur) and *Si* (silicone). The second letter is always lower case, to avoid mistakes. If one does not follow these conventions, great confusion results. *SI* is not silicone (Si), but sulfur and iodine. Some symbols derive from their Latin names. Examples: Pb for lead (from *plumbum,* whence "plumber"), Au for gold (from *aurum*), Na for sodium (from *natrium*). By international agreement, all countries use the same symbols.

The seven most common elements in nature, in order of abundance, are: oxygen, silicone, aluminum, iron, calcium, potassium, and sodium. The six most abundant elements in the human body are: oxygen, carbon, hydrogen, nitrogen, calcium, and phosphorus.

Procedure:

1. Select three elements.
2. Prepare a brief written report on these elements. Tell who discovered the element, when and where it was discovered and what its uses are. Add, if possible, how the element is processed to its pure form and where it is found.

4.2 ATOMIC STRUCTURE

Purpose: The purpose of this activity is to introduce subatomic particles.

Information: The atom is the smallest unbreakable particle of matter if the material is to keep its identity. Atoms are formed by even smaller subatomic particles. The main ones are *protons, neutrons,* and *electrons.* If you split an atom into its subatomic particles, you have subatomic particles and no identifiable matter.

All elements are formed by these same three particles. The various elements are different because each has unique numbers of these particles. Protons are positively charged in the center, or *nucleus*, of the atom. Neutrons are heavy, neutral particles in the nucleus of the atom. Electrons are light, negatively charged particles orbiting the nucleus, similar to satellites moving around a planet. If one assigns the relative mass of 1 to an electron, here are the approximate relative masses of the subatomic particles:

PARTICLE	RELATIVE MASSES	LOCATION	CHARGE
Electron	1	Orbit	Negative −
Proton	1843	Nucleus	Positive +
Neutron	1844	Nucleus	Neutral 0

One can see that protons and neutrons weigh almost 2000 times as much as an electron. From a viewpoint of mass, electrons are almost negligible. From a viewpoint of energy, they are critical parts of the atom. The atom has a nucleus with an enormous mass compared to that of its orbiting electrons.

SAMPLE BOHR DIAGRAM OF AN ATOM

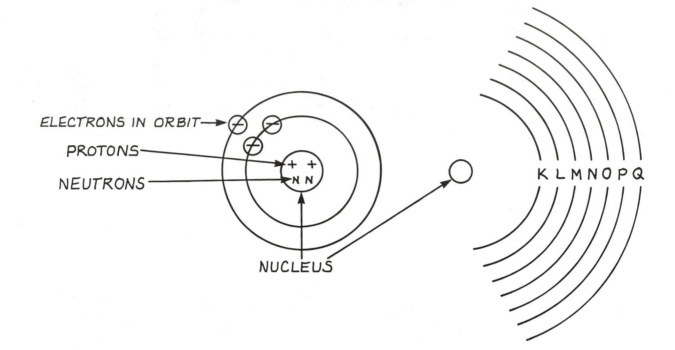

ELECTRONS IN ORBIT

PROTONS

NEUTRONS

NUCLEUS

K L M N O P Q

According to Einstein, the energy of an atom is in its nucleus. A nucleus is a small and heavy mass of material-energy, surrounded by negative charges which are far out in orbit. If the nucleus of an atom were represented by a tennis ball, the nearest orbital electron would be a basketball court away. The orbital shells, where electrons fly at the speed of light (186,000 miles/second or 300,000 kilometers/second), are well-defined. However one never knows where in a shell electrons are. The shells are named with letters of the alphabet, starting with *K, L, M, N,* and so on. *K* is the innermost shell.

Activity: On a sheet of paper draw the diagram of an atom and show where atomic particles are located. Label these particles.

Name _____ Date _____ Period _____

4.3 MIXTURES

Purpose: The purpose of this information is to introduce mixtures.

Information: Mixtures are nothing more than a mixing of atoms, molecules, or compounds without the formation of electrical bonds. Eventually, with some effort, you can separate mixtures. Each member of a mixture maintains its own properties. In a mixture of salt and water, the water evaporates to leave behind the salt. Since all materials are solid, liquid, or gaseous, there is no end to the number of combinations of mixtures that can be made.

	GAS	LIQUID	SOLID
GAS	air	soda pop spray paint shaving cream	plastic foam bread styrofoam
LIQUID	mist rain beaten eggs	tea and milk coffee and milk paintthinner	sugar in tea salt in sauce medicine in H_2O
SOLID	smog smoke dust	tomato sauce salad dressing cheese	plastics book pages watches

Procedure: Write a list of at least 15 substances you consider mixtures and explain why.

_____

_____

_____

_____

_____

_____

_____

_____

4.4 COMPOUNDS AND MOLECULES

Purpose: The purpose of this investigation is to provide students with an understanding of compounds and molecules.

Information: Compounds have their atoms interlocked electrically, forming molecules which have wholly new properties from their parent atoms. Ordinary kitchen salt is NaCl (sodium chloride), made up of sodium (Na) and chlorine (Cl). Sodium is an unstable, flammable, silvery metal. Chlorine is a nauseating and poisonous green gas. Salt, their chemical compound (union), is a white crystal used in ordinary human diets.

Chemicals make up everything around you. All molecules of matter are made up of at least two or more of the same or different elements. O_2, the oxygen molecule, is made up of two oxygen atoms. Aluminum tarnish is written Al_2O_3. It means that this molecule of aluminum oxide is made up of two atoms of aluminum and three of oxygen.

The number written below the line is a *subscript*. It tells how many atoms of that element are present. In NaCl (sodium chloride), there is one Na and one Cl atom. You use subscripts only when two or more atoms of the same kind and radicals are used in compounds. Radicals are groups of atoms that transfer together, unbroken, and cannot exist by themselves. Compounds are combinations of two or more of the same or different elements.

Procedure:

1. On a separate sheet of paper, copy the following names and formulas of compounds.
2. Write down the numbers of each atom. Find out what each symbol stands for.

Compounds:

1. Baking soda (sodium bicarbonate)—$NaHCO_3$
2. Silver tarnish—Ag_2O
3. Iron oxide (rust)—Fe_2O_3
4. Vinegar—$HC_2H_3O_2$
5. Water—H_2O
6. Sugar—$C_{12}H_{22}O_{11}$
7. Iodine—I_2
8. Copper oxide—CuO
9. Carbon dioxide—CO_2

Information: Molecules are two or more of the same or different atoms bound together by electrical charges. Molecules can be broken down into their individual atoms, providing energy is applied. Molecules combine to form new compounds. This process can either require or give off energy. If the process needs heat energy it is *endothermic;* if it gives off heat energy it is *exothermic.*

Questions:

1. How many atoms are there in iron oxide (rust)?_____

2. Which of the above compounds is only one element?_____

3. What do water and rust have in common?_____

4. Which molecule has the largest number of atoms?_____

5. What do iron rust, copper oxide, and silver tarnish have in common?_____

_____

_____

Information: Whenever a metal unites with oxygen, oxidation takes place. Examples: copper oxide, aluminum oxide. In the case of iron, we use the word *rust* to mean iron oxide. With silver we say it has tarnished when we know that it has oxidized. Fire is oxidation, where the process of oxidation takes place rapidly. Explosions are extremely rapid oxidations. Your body uses oxygen to burn and process food. The speed or rate of food oxidation in the body is called *metabolism.*

4.5 CHEMICAL AND PHYSICAL CHANGES

(Partner activity)

Purpose: The purpose of this investigation is to introduce physical and chemical changes.

Information: A *physical change* forms no new substance. It merely changes the shape, form, or state of a material. Examples: boiling of water, evaporating water, melting of ice, freezing of water, tearing a piece of paper, crushing a can, moving an object.

A *chemical change* forms one or more new substances. Some familiar chemical changes: burning of gas, oil, coal, or paper; the souring of milk; the rusting of iron. Chemical changes usually involve heat, burning, or some interaction with energy. Chemical changes are either endothermic (absorb heat energy) or exothermic (give off heat energy). A chemical change makes new compounds. There are five ways to tell if a chemical reaction has occurred:

1. A change in color appears.
2. Gases are given off.
3. A new substance forms.
4. Heat is given off (an exothermic reaction).
5. Heat is required (an endothermic reaction).

Sometimes you can see only one of these, sometimes several.

Equipment: Teaspoon, sugar, baby-food jar, aluminum foil, water, matches, candle

Procedure:

1. Prepare data form as shown.
2. Mark the appropriate answer with a check.
3. In the following investigations, identify each change as being either chemical or physical.
   a. Mix a small amount of sugar with water in a jar.
   b. Make a small aluminum trough with the aluminum foil.
   c. Scoop up with a teaspoon a drop of water
   d. Place the water drop in the trough.
   e. Strike a match.
   f. Light a candle.
   g. Carefully boil the water.
   h. Evaporate the water.
   i. Place a little sugar on the trough and burn it.
   j. Tear a piece of paper.
   k. Throw it away.
   l. Take a small amount of flour and place in the trough and burn in.
   m. Place a tiny piece of paper in the trough and char it.
   n. Place a small amount of starch in your hand and mix with a drop or two of water to make a dough.

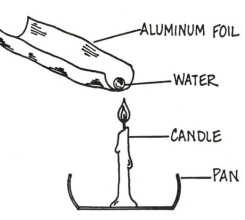

ALUMINUM FOIL

WATER

CANDLE

PAN

110

o. Place a teaspoon of baking soda in a baby food jar, add some water, and stir it.
p. Add a few drops of water to the baby food jar.
q. Add a few drops of vinegar.

Sample Data Setup:

#	Chemical change	Physical change
1.		
2.		
3.		
4.		

add as many lines as needed.

Questions: Identify the following as chemical or physical changes:

1. Burning of sugar or marshmallows
2. Sharpening a pencil
3. Cooking an egg
4. Baking cookies
5. Adding sugar to tea
6. Toasting a slice of bread
7. Mixing flour and water
8. Mixing ice cream with soda
9. Sifting flour
10. Burning coals in a barbecue

4.6 EVIDENCE FOR OXYGEN

Purpose: The purpose of this investigation is to prove the presence of oxygen in air.

Information: Oxygen is a gas which is part of air and the human body. Air is a mixture of many gases, including about 21 percent oxygen.

Equipment: Test tube, a small piece of steel wool, baby-food jar, masking tape, paper clip, ruler.

Procedure 2:

1. Wash steel wool with detergent to remove oil film placed by manufacturer.
2. Insert the steel wool into the bottom of the test tube and hold in place with a paper clip.
3. Fill baby-food jar with about 3 cm to 5 cm of water.
4. Invert test tube and place into baby food jar.
5. Label jar and measure height of air column for several days.

Sample Data Setup:

Date	Height of air (cm)

Questions:

1. Did the height of the air column in the test tube change?_____
2. By how much?_____
3. Why did it change? _____

   _____

   _____

4. Did the steel wool change color? What is this called?_____
5. Why do manufacturers oil the steel wool? Why are most metals painted?

   _____

4.7 PERCENTAGE OF OXYGEN IN AIR

(Partner activity)

Purpose: The purpose of this investigation is to verify that oxygen makes up 21 percent of the air and to learn about the processes of oxidation and explosion.

Caution: Follow all federal, state, and local safety rules.

Information: In this investigation you measure the approximate percentage of oxygen in air. Air is a mixture of many gases, including nitrogen—over 70 percent, oxygen—21 percent, water vapor, carbon dioxide, helium, neon, hydrogen, ozone, and others.

Equipment: Small candle, food coloring, shallow pan filled with 4 cm to 5 cm of water, several glass jars (graduated cylinders or peanut butter jars, for example).

Procedure:

1. Affix the candle to the bottom of the water dish. Use of a few drops of hot wax.
2. Place a piece of masking tape on the side of the jar, from the top to the bottom. You can mark it during the activity, then measure it.
3. Fill dish with water and color the water. Place a jar on top of the candle (do not light it yet), and measure the height of air in the jar.
4. Light the candle, cover it with the jar, and see what happens. Measure the height of air in the jar. Write down all observations.
5. Repeat this activity several times.

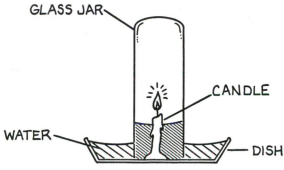

Sample Data Setup:

A	B	C
height of air in jar (cm) at start	measured height of air in jar (cm) after burn	Difference $(B-A=C)$ air in jar (cm) after burn

Average: (A) _____ cm   (B) _____ cm   (C) _____ cm

Questions:

1. When you first cover the lit candle, did you see bubbles escaping? What causes them?

_____

_____

2. Why does water rise in the jar?_____

_____

_____

Calculations: To find out the percentage of oxygen in air, you divide the *difference* in the air columns (heights before and after the burn) by the *total* air column height before the burn. Then multiply the answer by 100. Let A = 17.4 cm (column before burn), B = 13.5 cm (column after burn).

1. Subtract: $B - A$. Call this difference C.
2. Divide C by the value of A and multiply by 100. This will give you the percentage of oxygen. Example: $(B - A = C)$, $17.4 - 13.5 = 3.9$; $(C/A) \times 100$, $3.9/17.4 \times 100 = 22.4\%$. This is fairly close to 21%.
3. Show all your calculations on a clean sheet of paper. Label it *calculations* and turn it in with your lab report.

4.8 ACIDS, BASES, AND LITMUS PAPER

Purpose: The purpose of this investigation is to introduce acids and bases, using chemical indicators.

Information: All compounds in nature are one of the following: acids, bases, or salts. Acids are sour, and bases are bitter to the taste. Acids have hydrogen (H) ions, while bases have hydroxide (OH) molecules. Both react readily and are caustic (cause damage) to skin. When acids mix with bases in the right amounts, they are both neutralized. The result of mixing acids and bases is water and salts.

In this activity you test substances to determine whether they are acids, bases, or neutral. *Alkali* is a synonym for base. Scientists measure the acidity or alkalinity of substances by using the pH scale.

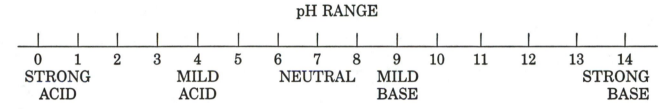

pH RANGE

| 0 | 1 | 2 | 3 | 4 | 5 | 6 | 7 | 8 | 9 | 10 | 11 | 12 | 13 | 14 |

STRONG ACID   MILD ACID   NEUTRAL   MILD BASE   STRONG BASE

When pharmacists make products that come in contact with the human body, they match the product's pH with that of the body. Otherwise the products sting on contact. This applies to facial cream, lipstick, hair shampoo, eye washes, etc. The process of matching the pH of a chemical to where you use it is called *buffering*.

Examples of pH of substances you handle daily: lemon juice, 2.2 pH; vinegar, about 3 pH; cow's milk, 6.3–6.6 pH; baking soda (diluted), 8.4 pH; ammonia in water (diluted), 11.11 pH; TSP cleaner (trisodium phosphate diluted), 12.4 pH.

Indicators are used to find out the pH of substances. They change color when placed in contact with acids, bases, or neutral materials. Many indicators are organic, that is, they contain carbon. There are indicators to show pH, the hardness of water, the presence of certain chemicals, and so forth.

Equipment: Four strips of litmus paper, two blue and two red. Two glasses, one containing an acid (vinegar and water), the other a base (baking soda and water). Clearly label both.

© 1991 by The Center for Applied Research in Education.

Procedure:

1. Test the red and the blue litmus papers, by dipping them in the acid solution. An indicator usually changes its color right away. If both litmus papers show no change, that indicates a neutral substance.
2. Record the results.
3. Repeat with the base.
4. Throw away the test strips in a disposal dish.

DO NOT HANDLE CHEMICAL SUBSTANCES. RINSE HANDS WITH WATER IF YOU COME IN CONTACT WITH ANY CHEMICALS!

Sample Data Setup:

LITMUS PAPER TESTS

	RED	BLUE
ACID		
BASE		

Questions:

1. Which color litmus is a good indicator for an acid?_____
2. Which color litmus is a good indicator for a base?_____
3. Make up and write a rhyme to remember it!

_____

_____

4.9 LITMUS PAPER AND HYDRION

Purpose: The purpose of this investigation is to use litmus paper and other indicators to find whether substances are acid, neutral, or basic.

Information: Litmus paper is an organic indicator. You use it to find out whether a substance is an acid, a base, or if it is neutral. Red litmus turns blue in the presence of bases, and blue litmus turns red in the presence of acids. If both strips of litmus do not change color, this means that the substance is neutral. Hydrion is an indicator which turns yellow with acids, light green with neutral substances and deep blue with bases.

Equipment: Litmus strips both blue and red, one of each for every sample. One hydrion test strip for each sample.

Procedure:

1. Prepare a data sheet.
2. Test each available substance with both litmus red and blue to find out whether it is basic, acid, or neutral. Repeat investigation using hydrion indicator. Write your outcome in the box labeled *Results*. At the end of this activity, the instructor will reveal the contents of the samples which you tested.

Sample Data Setup:

#	Material	Litmus Red Color	Litmus Blue Color	Results	Hydrion Color	Results
1						
2						
3						
4						
5						
6						
7						

Other Indicators (Homework Assignment):

Information: Chemical indicators are used in laboratories, industries, businesses, schools, and homes. Methylene red is used for swimming pool pH tests. Phenolphthalein, bromothymol blue, and many other indicators are used daily to perform chemical tests.

Homework: Test at least ten common substances you find around the house. Examples: milk, tea, soap, detergent, shampoo, ammonia, window cleaner, mouthwash, toothpaste, face cream. Set up a data box to show whether these substances are acid, neutral, or basic. Get hydrion indicator lines on your paper from your teacher.

Caution: Keep the hydrion dry until use.

Procedure:

1. Copy this information on a sheet: Yellow = Acid; Pale Green = Neutral; Blue = Base.
2. Mark below each hydrion mark a consecutive number.
3. When testing, read and record results immediately.
4. Moisten solids which you wish to test, unless they have their own juices, like fruits and vegetables.

Safety Note: Do not test swimming pool acid or chemicals used to clean drains. Do not mix any chemicals prior to test. You could be severely hurt! If you test laundry bleach, first ask your parents to dilute a teaspoon of bleach in 3 or 4 oz. of water. Do not test oils or honey, for they will make a big mess on your test paper. Expect to be surprised by some of the results!

Sample Data Setup:

#	Material Tested	Color	Result
1			
2			
3			

Add as many additional lines as needed.

Questions:

1. Are there any other substances that you wish to test? Name them.

_____

_____

4.10 BROMOTHYMOL BLUE VS. BAD BREATH

Purpose: The purpose of this investigation is to become familiar with the organic indicator *bromothymol blue.*

Information: An indicator is a substance which changes color in the presence of other chemicals, because it reacts with them. The color change is usually immediate and the new color is usually entirely different. We use this property of indicators to find out whether substances are acid, whether they contain starch, whether water is hard, etc. The word *organic* means that the substance contains the element carbon. This is important, for it suggests that the substance comes from formerly living materials, such as plants or animals.

Equipment: Glass or plastic container, tap water, drinking straw, bromothymol blue

Procedure:

1. Teacher places one drop of bromothymol blue in a glass of water.
2. Bubble your breath through the soda straw until the color of the liquid changes.

Questions:

1. Did the liquid change color? Explain it.

_____

_____

2. If you want to change the liquid into its original color, what needs to be added? Explain.

_____

_____

4.11 TESTING FOR STARCHES WITH IODINE

Purpose: The purpose of this investigation is to use an indicator to check for the presence of starch.

Information: Starch is a common substance used for foods and by industry as filler or stiffener. You probably have made glue with flour. It works well because flour has starch in it. Starch is also used to stiffen clothes before ironing. Water soluble starch is formed by many molecules of sugar. Each connects to the next in long chains, like beads in a necklace. Non-water-soluble starch molecules are strung together like many twisted beads, going out in all directions. Starch is a carbohydrate, and it breaks down during digestion into short segments of glucose (a form of sugar) to be used for energy. Iodine turns dark blue or black in the presence of starch, making it an ideal indicator for starches.

Equipment: Apron (optional), iodine (diluted) in a dropper dispenser, flour, baby powder, sugar, confectioner's sugar, baking powder, baking soda, salt, starch, a small piece of paper towel, a piece of paper used by students and other substances to be tested. Use Hypo for stain removal in case of accident. (Hypo is for sodium hyposulphite. It is used to fix and harden photographic films and papers.)

Procedure:

1. Take a paper plate and, with a pencil, section it into wedges.
2. Label these with the names of the substances you test.
3. Also test the paper plate itself.
4. Shape all materials into little mounds with shallow craters on top, like tiny volcanoes. Place the drop of iodine in the crater. Be careful to have some material between the bottom of the hole and the plate. Otherwise, you test the plate. Once is enough.

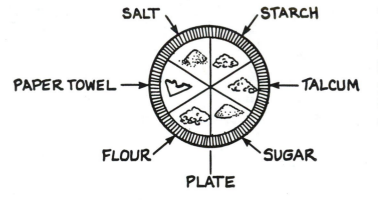

© 1991 by The Center for Applied Research in Education.

Sample Data Setup:

Substance tested	prediction starch Y/N	test color w/iodine	Does it have starch? Y/N

Use as many lines as needed.

Questions:

1. List all substances tested which contain starch.

_____

_____

2. What other substances would you like to test?

_____

_____

4.12 WATER HARDNESS

(Partner activity)

Purpose: The purpose of this activity is to compare the relative hardness of three samples of water.

Information: Most people are aware of rings around bathtubs and white deposits on shower doors. If one boils water in a pot, rings appear on the inside. Minerals dissolved in tap water cause these white material deposits. Iron, calcium, magnesium, and other minerals give water its flavor.

Water with a high mineral content is *hard water*. Distilled and demineralized water contain no minerals. When soap is added to hard water it does not make suds. It forms residues, and you must use large amounts to clean with it. Water where sodium has replaced calcium is *soft water*. Industry and many homes use water softeners. These are water treatment appliances which replace hard water's calcium with sodium. While soft water does not produce soap residues and needs little detergent, the sodium can be a problem for people with high blood pressure.

Equipment: Three test tubes with support rack, liquid soap (not a detergent) in dropper-bottle or bottle with separate dropper, several water samples (hard water, soft water, river water, sea water, etc.), ruler

Procedure:

1. Prediction: Rank on the data chart your water samples from high to low mineral content. (1 = low, 2 = middle, 3 = high) to see how informed you are.
2. Fill all tubes to equal heights (half full) with water from each sample.
3. Using the dropper, add one drop of liquid soap to one sample. Shake it for 30 seconds exactly and look.
4. If no suds appear, stop, add another drop of soap, and repeat the shaking.
5. When suds appear, let them stand for 3 minutes. If the suds appear to collapse, do not wait the full 3 minutes, add another drop of soap and repeat.
6. If the suds stand without a change for three minutes, your test of the particular sample is finished. Record immediately the number of drops of soap used.
7. Now repeat steps 3 through 6 with the other water samples.

Sample Data Setup:

Rating scale: 1 = low mineral content
2 = medium mineral content
3 = high mineral content

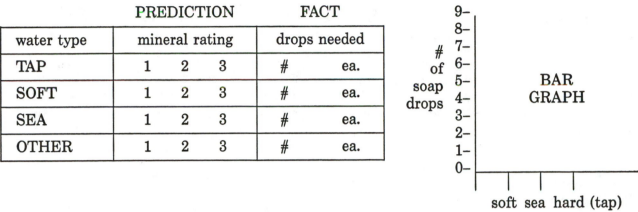

water type	mineral rating			drops needed	
TAP	1	2	3	#	ea.
SOFT	1	2	3	#	ea.
SEA	1	2	3	#	ea.
OTHER	1	2	3	#	ea.

Use as many lines as needed.

Questions:

1. Which water needed the most soap drops? Why?

_____

_____

2. Which water required the least number of soap drops? Why?

_____

_____

3. After the suds were in place, did the tap water look different from the soft water? Explain.

_____

_____

4. Can you think of applications for the use of soft water?

_____

_____

Name _____ Date _____ Period _____

4.13 UNKNOWN POWDERS: VARIABLES

(Partner activity)

Purpose: The purpose of this investigation is to identify similar looking substances by observing and comparing their properties when exposed to several reagents.

Information: You will sharpen your observation and science skills with this investigation. You will receive six white powders and three reagents: water, iodine, and vinegar. A *reagent* is a chemical substance which produces a specific reaction when in contact with other materials. The powders are in jars marked with numbers.

Equipment: Labeled jars with powders, water, iodine, vinegar, plastic teaspoon, toothpicks, paper plates.

Procedure 1:

1. Divide your plate into six wedges. Label each wedge from 1 to 6.
2. Place in each slice about 1/4 teaspoon of each powder. Make certain that the numbers on the plate match the numbers on the jars.
3. Form a tiny mound with each powder and shape the top like a volcano crater.
4. Use the *same* reagent on each powder.
5. Place one drop of reagent into the small hole on top of a mound.
6. Stir powder with a toothpick.
7. Record the reactions, if any.
8. Continue until you have tested each sample.
9. Throw away the plate.
10. Get a second plate and repeat steps 1 through 9 using the next reagent.
11. Get a third plate and repeat steps 1 through 9 with the last reagent.

Procedure 2:

1. Repeat steps 1 through 11 using jars 7 through 12.

MOUND

Sample Data Setup:

unknown powder	REAGENTS		
	WATER	VINEGAR	IODINE
1			
2			
3			
4			
5			
6			
7			
8			
9			
10			
11			
12			

Questions:

1. Which powders turned black?

_____

2. Which powders bubbled?

_____

3. Which powders did not mix with water?

_____

4. If two powders bubble, how can you distinguish between them?

_____

_____

_____

5. Unknown powder 7 is most alike to which sample? Why?

_____

_____

_____

6. Unknown powder 8 is most alike to which sample? Why?

_____

_____

_____

7. Write a description for all powders you have used and tell how they react with each different reagent.

_____

_____

_____

_____

_____

_____

_____

_____

4.14 NUCLEAR ENERGY: FISSION

Purpose: The purpose of this investigation is to provide a model for nuclear fission.

Information: The nucleus of an atom contains protons (positively charged particles) which repel each other with great force. Nuclear energy, a very strong force, keeps the protons and neutrons together in the nucleus. Neutrons, heavy bundles of matter or neutral energy, are electrically neutral. They neither attract nor repel protons or electrons. If somehow one manages to insert an extra neutron into the nucleus of the atom, then the nucleus becomes unstable. The internal forces become greater than those of the binding nuclear energy. The atom splits up, producing heat, particles, and radiation.

Radiation is radioactivity. There are three varieties:

1. *Alpha* particles are the weakest and can be stopped by human skin or a sheet of paper.
2. *Beta* particles are more intense and can penetrate the skin.
3. *Gamma* radiation will go through people and can damage human body cells. If an atom of uranium^{238} splits, it frees 146 neutrons which shoot out and by chance hit another U^{238}, which in turn frees 146 more neutrons. Scientists expect at least two of these neutrons to split additional atoms of U^{238}.

The reaction goes like this:
$$1 \rightarrow 2 \rightarrow 4 \rightarrow 8 \rightarrow 16 \rightarrow 32 \rightarrow 64 \rightarrow 128 \rightarrow 256 \rightarrow 512 \rightarrow 1024 \rightarrow 2048 \rightarrow 4096.$$
If you continue twenty-one more times, you have 8,589,934,592 fissions.
Fission is the scientific word for nuclear splitting. The amount of nuclear energy released from fission is very large. Nuclear power plants make electricity with uranium at a low concentration, around 3 percent. To have a nuclear explosion, you need a concentration of uranium greater than 92 percent. This is very difficult to accomplish. It is done with a process called gaseous diffusion. In the United States, the government operates the only factory which produces this weapons grade uranium. In nuclear power plants, fission is continuously monitored with capability of shutdown in less than one tenth of a second. Nuclear power plants have safety equipment to prevent accidents. Accidents have occurred, however, usually due to human error, not machine failure.

Procedure 1:   A way to stimulate a chain reaction is as follows: Everyone brings to school their dominoes. Set up a domino chain, and by tripping only one, a hundred or more tumble. Follow the pattern at right: Use playing cards as an alternative.

Procedure 2:

Another way to illustrate a chain reaction is by using mousetraps, 4 inches by 1¾ inches. You will need about twenty-one traps per square foot.

1. In a corner of the room set up an area about 3 to 5 feet square, with mousetraps filling this area completely.
2. Load each mousetrap and place a ping pong ball on top.
3. When you throw one ping pong ball on this setup, the chain reaction simulation starts—no mice needed!

4.15 AN OIL-DROP MODEL OF SPLITTING AN ATOM

Purpose: The purpose of this investigation is to provide a visual model for the splitting of an atom.

Information: Many scientists suggest that splitting an atom resembles the behavior of a drop of liquid when it breaks up into smaller droplets. *Nuclear fission* is the splitting of an atom. *Nuclear fusion* is the process by which four atoms of hydrogen combine to form one atom of helium, with a large release of energy. This process takes place only in the presence of incredibly high temperatures, as on the surface of the sun. People hope to harness and use this form of energy by the 21st century.

Equipment: An 8 to 12 oz. glass, small bottle of rubbing alcohol, a few ounces of cooking oil, water, paper towels, teaspoon, eye dropper

Procedure:

1. Fill glass about half full with rubbing alcohol.
2. Add enough water to fill to about three-quarters full.
3. Stir the mixture with a teaspoon.
4. Fill eyedropper with salad oil and insert dropper into the water-alcohol mixture about half way up from bottom. Slowly release the oil into the liquid.
5. If the large oil drop floats on or near the surface, add a little more alcohol with a teaspoon, away from the large oil drop.
6. If the large oil drop is near the bottom, gently add a bit more water to the mixture with the teaspoon.
7. Continue with steps 5 or 6 until the large oil drop floats in the middle of the glass.
8. Note the near spherical shape of the large oil drop. It is a model of atomic forces.
9. Using the handle of the teaspoon, cautiously force the drop apart. First the drop will appear to bulge, then it will split apart into two or more round oil drops. The oil-drop "atom" will have split into two or more smaller "atoms."

Observe that before the oil drop appears to split, it swells and changes shape. When atoms split, they too resist splitting until, at the critical moment, they appear to be deformed by outside action.

© 1991 by The Center for Applied Research in Education.

Questions:

1. Name two nuclear processes that release energy.

   _____

2. Describe how one works.

   _____

   _____

3. Describe how the other one works.

   _____

   _____

   _____

   _____

ENERGY

TEACHER'S SECTION

In this section students will learn and perform activities about:

1. Potential and kinetic energy
2. Conduction of heat
3. Energy form changes
4. Convection
5. Electromagnetic spectrum
6. Radiation
7. Wave terminology and radiation
8. Invisible energy
9. Calories in food

5.1 POTENTIAL AND KINETIC ENERGY

Vocabulary: Energy, friction, mechanical, gravitational

5.2 CONDUCTIVITY OF HEAT

Caution: Supervise use of candles carefully. If they are illegal in your classroom, substitute Bunsen burners.

Answers:

1-4. The molecular structure of metals is regular and molecules are close to each other. This regularity makes it easy to transfer vibrations. In wood and air, molecules are unevenly spaced and farther apart.

5.3 ENERGY TRANSFER: WHIPPED CREAM

I suggest that two or three teams perform this activity for the entire class. Keep cream as cold as possible. Freeze if possible up to one hour before use. Ask students to bring to class from home the various beaters and mixers. Wash all equipment before use and between activities. Students need to use teaspoons to eat whipped cream.

Caution: If any student has high cholesterol, ask them not to taste the whipped cream.

Note: This is an exercise in energy form exchange. Student predictions will favor the electric model. The manual method is the usual winner because it transfers the energy most efficiently. The sequence of energy changes:

1. Hand operated beaters
   Solar energy → food energy → personal energy → mechanical energy → heat energy.
2. Electrical beaters
   Chemical energy (oil) → electrical energy → magnetic energy → mechanical energy → heat energy. The whipping cream will warm up during the activity due to room temperature and the addition of the mechanical energy of whipping.

5.4 CONVECTION: WINDMILLS AND ROTORS

Caution: If you are unable to use candles because of safety rules in Procedure 1 go directly to step 5 (skip 1 through 4).
Applications of hot gases:

1. Hot air balloons
2. Pistons in an engine move due to gas expansion.

The warmest spot in the class is near the ceiling. Convection is responsible for moving hot air and water masses up and cold air and water masses down. This accounts for winds, tornadoes, hurricanes, and the entire cycle of weather as well as marine currents like the Gulf Stream.

DEMONSTRATION: CONVECTION OF LIQUIDS

This is one of the finest activities you can do. It illustrates many concepts. It shows convection, thermal layers, thermoclines, etc.

Information: Convection explains why hot water rises and cold water sinks. Hot water is lighter because its molecules are farther apart and it has a mass less than the same volume of cold water. (Cold water has more molecules because they are closer together).

Equipment: 3 or 4 identical baby-food jars with lid, two 6 cm glass tubes (about the size of an eye dropper) or use plastic eye droppers, florist's clay, food coloring, large jar or fish tank (best), hot and cold water. If an electric hot melt glue gun is available in place of florist's clay, it is preferable. The lids you will prepare can be moved from jar to jar.

Procedure to Prepare Apparatus:

1. Take a baby-food jar lid and pierce two holes in it.
2. Fit into one hole a small glass tube or plastic tube of an eye dropper (with both ends cut off to widest point). Fit 90 percent of the tube inside the jar.

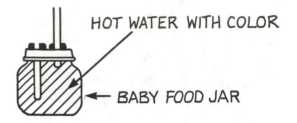

3. Seal the tube with florist's clay or hot melt glue on the upper side of the lid.
4. Place the other glass tube 90 percent outside the lid and seal on the underside of the lid with clay or glue.
5. Prepare a couple of extra lids. It is a great present for visiting teachers.

Procedure for Activity:

1. Fill jar with warm water and let it stand for a few minutes, to allow for heat expansion.
2. Refill jar with hot water and add several drops of food coloring.
3. Cover jar with the lid you prepared and cover the upper tube with your finger while lowering jar into fishtank or large jar full of tap water.
4. Observe water convection. The effect is even more pronounced if you prepare a second jar filled with tap water, using a different color. Demonstrate its lack of convection by placing it in the same fishtank or jar. Notice how the warm water forms a layer on top of the fishtank, illustrating an inversion layer. This can introduce a superb weather lesson.

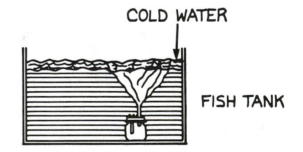

5. Repeat activity several times using different colors. The top of the tank will have different colored layers. In oceanography these are called *thermoclines*, layers of water at the same temperature.

Convection Currents: Nowadays it is popular and energy efficient to cool homes by having an opening with a fan at the top of the house, to allow rising hot air to escape and fresh air to enter and replace it. This is convection at its best with a little mechanical help. Old cars used to depend on thermo-syphoning (convection) to move their cooling water in and out of the radiator. This was before the development of water pumps, and it explains the many car overheating episodes in old-time movies. In modern days, thermo-syphoning is still used to cool the turbine oil in some cars with somewhat similar consequences. Also note that humans lose 80 percent or more of their body heat through their heads. It is therefore very important to wear a hat in cold weather.

Vocabulary: Convection, fluid, wind, tornadoes

5.5 ELECTROMAGNETIC SPECTRUM, RADIATION, AND INFRARED RAYS

One of the objectives here is to learn that radiation is not a taboo word. Most students associate it with deadly rays. You should display an inexpensive spectrum chart. Here you have the electromagnetic band at times arranged with the shortest waves on the left and the longest waves at the right. *The type of any energy is determined by its wavelength.* Among the shortest wavelengths are the radioactive rays. Ultraviolet rays are longer, followed by visible light, infrared rays (heat), radio waves, and sound waves.

Spectrum charts are useful for many other applications. They may show some spectra of elements, the Doppler shift of the sun, Fraunhofer lines, spectroscopes, etc.

DEMONSTRATION: TRAVELING WAVES

Equipment: Slinky™ (the metal model is preferable), two students

Procedure:

1. Take a Slinky™ rolled spring coil and have two students stretch it about 8 m (24 ft.) apart in the air.
2. Pluck it so everyone can observe the action of traveling waves. The wave eventually stops because it meets friction in both the metal and the air molecules. Do not overstretch it. You can even hear interesting sounds, if you place the end of it near an ear.

DEMONSTRATION: WAVE TERMINOLOGY

Information: Waves of energy come in many forms. Some travel horizontally, some vertically. Some waves are caused by compression (push together) forces.

Equipment: Slinky™

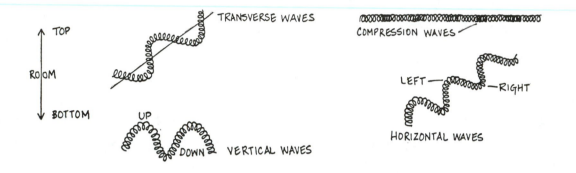

Procedure:

1. Two students about 7 m (23 ft.) apart hold a Slinky™.
2. Pluck Slinky™ laterally. This will be a model of a (lateral/horizontal) transverse wave. A transverse wave travels at an angle to a medium.
3. Push the Slinky™ suddenly forward with much energy, as if to bring it together, providing a model of compression waves. This is a model of a longitudinal wave. A longitudinal wave travels along the same direction as the medium.
4. Pull the Slinky™ briskly up and down to generate vertical waves.

Note: TV waves travel in the horizontal plane. This accounts for why all TV antenna elements are horizontal (look at the antennae on roofs).

Vocabulary: Radiation, propagation, electromagnetic spectrum, infrared

5.6 CALORIES IN FOOD

You may want to use other foods to test. Choose oily nuts, etc., because the samples are small, burn well, and are easy to manage.

Answers:

1. 500 calories
2. 1000 calories
3. 25,000 calories
4. 100,000 calories
5. 1.0 calorie
6. 55,000 calories, 55 food calories
   Regular peanuts usually have 1000 calories or one food calorie.
7. a. Neither. Both have the same temperature.
   b. The pot has more stored heat energy. The pot is full of water and it takes many more calories to heat it than a tablespoon.

Vocabulary: Combustion, calorimeter, calorie

5.1 POTENTIAL AND KINETIC ENERGY

Purpose: The purpose of this activity is to define energy and to learn more about its many forms.

Information: *Energy* is the ability to do work and to move masses. Energy comes in many forms: solar, gravitational, wind, nuclear, electrical, magnetic, chemical, heat, light, sound, radioactive, cosmic rays, radio waves, microwaves, and more. Light is one form of energy which is visible. Other forms of energy are invisible. You feel heat and hear sound. Ultraviolet energy darkens your skin. X-rays go through you. Radio waves carry sound and music and pictures. These five forms of energy are invisible.

Energy exists in two modes, potential and kinetic. *Potential energy* is energy at rest and available, for example:

- Electricity before you turn on the switch to light a room
- Gasoline in a car's tank
- Wax in a candle

- Wood log for a fireplace
- Battery in a flashlight
- Food before you eat it

Kinetic energy is energy in motion, like the energy of motion of a car. Potential energy, like gasoline in a car tank, changes into chemical energy when it burns inside the motor. This burning process provides heat and pressure energies which move many parts of the engine (mechanical energy). The moving engine produces electricity by turning the alternator. The electricity in turn blows the horn (sound), lights the headlights (light), turns the fans for the air conditioner and heater (mechanical energy), heats the cigarette lighter (heat).

Mechanical energy involves real moving parts like wheels, levers, gears, cranks, pulleys, etc. *Friction* is a force (resistance) that opposes the motion of one surface past another.

Energy changes from one form into another at a price. We lose some energy in the process. This lost energy is not useful to us. It is a waste most of the time. For instance: You boil water. The heat energy under the kettle heats the water and the air around the kettle and has to replace the heat which the kettle loses to the cooler air around it. Transmission lines lose part of their electricity due to heat caused by the wire's internal resistance. Scientists and engineers try to reduce this loss by applying scientific principles and more efficient designs.

Equipment: Rubber band, paper clip, scissors, pin, pencil

Procedure 1:

1. Stretch a rubber band between two fingers ready to let it fly. When the band is stretched out, it has potential energy (stored energy), which it is ready to release.
2. Launch the rubber band (not at anybody!): it has kinetic or motion energy.

Procedure 2:

1. Place both hands against each other and rub briskly. Your hands should get warm. The rubbing friction of your hands is turned into heat energy.
2. Open a paper clip and bend it back and forth as if you are trying to break it.
3. After a while, place its center on your forehead and feel it. It should be warm—you have changed mechanical energy into heat energy.

Procedure 3:   Build a small paper pinwheel:

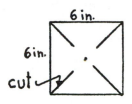

1. Cut a square piece of paper 12 cm to 15 cm on each side.
2. Draw the diagonals.
3. Cut half way from each corner to the center.
4. Fold each right corner to the center without creasing the paper.
5. Place a pin through it. Place a dime-size piece of paper behind it as a washer.
6. Stick the pin in the eraser of your pencil. You have a model windmill.
7. Blow on it or place it near an air duct.

Procedure 4:   We define energy as the ability to do work or move masses.

1. Look at the table.
2. Copy the examples and add several more.

ENERGY	OBJECT	RESULTS
wind	trees	bends or breaks
chemical	food	digestion
gravity	ball	falls down
light	plants	trees grow toward it
electrical	motor	moves washing machine parts
mechanical	wheelbarrow	moves dirt
heat	water	steam moves turbines (to make electricity)

5.2 CONDUCTIVITY OF HEAT

(Partner activity.)

Purpose: The purpose of this activity is to investigate the conductivity of heat by different materials.

Information: Heat is one of the many forms of energy radiated by the sun. If all heat is withdrawn from material objects, molecular action stops and objects shrink to almost nothing. Heat boils water, makes steam to generate electricity, expands bridges, and moves cars. This happens because heat energy speeds up the motion of molecules. With heat added, molecules vibrate further apart and faster.

When a metal is heated, its molecules vibrate and send their excess energy to other nearby molecules. This process is *conduction*. Since the molecules move faster and farther apart, the metal itself expands. You have seen the marker liquid rise in a thermometer: it expands because of heat energy. Some materials, like metals, are good heat conductors. Wood, glass, paper, air, and some plastics do not conduct heat well. Nonconductors are insulators. Ice chests use high density foam plastics to insulate foods. NASA Space Shuttles have their skins covered by special tiles to insulate them from fiery reentry heat. Water heaters have an internal layer of insulation to keep heat energy from escaping. People use clothes as insulators. Homes use many forms of insulation to keep heat in or to prevent heat from coming in.

Equipment: Candle or Bunsen burner, paper clip, aluminum foil

Procedure 1:   To check if air is a good conductor of heat:

1. Light a candle or Bunsen burner and carefully move your hand close to the flame, not any closer than 6 cm. Do not burn yourself! Since air is a good insulator, you should not feel any heat.
2. Move a straight wire, an opened paper clip, to about 6 cm from the flame.
3. Move part of the paper clip into the flame.
4. Repeat step 3 with a 2 cm × 4 cm strip of aluminum foil.

Questions:

1. Did the paper clip transmit heat from 6 cm?_____
2. Did the paper clip get hot when one end was in the flame?

_____

3. Did the foil transmit heat?_____

4. Why is air a poor conductor, while metals are good ones?

_____

_____

_____

Procedure 2:

1. Place a little water in a paper cup.
2. Heat the paper clip in the flame.
3. Quickly place the clip in water.

Questions:

1. What happened when you placed the hot paper clip in water?

_____

_____

_____

2. Why do people in Alaska and other cold places wear many layers of clothes?

_____

_____

_____

3. Why do people use down-filled comforters in winter?

_____

_____

_____

5.3 ENERGY TRANSFER: WHIPPED CREAM

CLASS ACTIVITY (Teams of three students)

Purpose: The purpose of this activity is to investigate which method of whipping cream transfers more energy to the cream in the least amount of time.

Information: This is a race to see how fast you can whip cream using three different ways:

1. With an old-fashioned hand beater
2. With a hand-crank beater
3. With an electric mixer

Equipment: Three mixing bowls, cold heavy cream, powdered sugar, hand beater, crank beater, electric mixer, dishwashing liquid, plastic teaspoons, clock on wall or three stopwatches.

Procedure:

1. Three students are picked as timekeepers.
2. Everyone prepares a data sheet.
3. Rate the order in which you think whipped cream will be ready.
4. Pour a small, identical amount (½ to 1 cup) of cream in each bowl, and add powdered sugar to taste.
5. At the signal the teams work hard to whip up the cream. The cream is whipped when it stands upright for one minute without liquid residue.

Sample Data Setup: Predictions for order of whipped cream
RATINGS SCALE: 1 = first to finish, 2 = second to finish, 3 = third to finish
MANUAL: 1 2 3; CRANK: 1 2 3; ELECTRIC: 1 2 3; (Circle numbers to show order of your prediction).

TRY #	TIME FOR METHOD		
	MANUAL	CRANK	ELECTRIC
1			
2			
3			
4			
5			

AVERAGES →

Questions:

1. Which method did you predict as the winner?

   _____

2. Which method won? Why do you think so?

   _____

   _____

   _____

3. What are the energy changes in this activity? Name them.

   _____

   _____

   _____

   _____

   _____

   _____

5.4 CONVECTION: WINDMILLS AND ROTORS

Purpose: The purpose of this activity is to examine the effects of convection, a property of heat.

Information: Heat energy from a burning gas can move masses. Air and water are poor conductors of heat, so heat moves around by a process called *convection*. In this process hot air or water (fluids) rise up while the cold air or water sink down.

Equipment: Two candles, two thermometers, compass, thread, paper clip, aluminum foil, aluminum plate (pie tin), ruler, scissors, shallow pan (for candles), candles
Alternative hot gas source: hot plate with boiling water in a beaker.

Caution: HANDLE YOURSELF CAREFULLY AROUND ANY HEAT SOURCE!

Procedure: Build a small paper heat engine:

1. On a piece of paper, mark a circle with a diameter about 10 cm to 15 cm.

2. Draw a spiral on it and cut it out.

3. Make a pinhole in its center.

4. Pass through the pinhole a small piece of thread about 15 cm to 20 cm long.

5. Tie a knot on one end and a paper clip on the other.

6. Hold the spiral 15 cm above a Bunsen burner flame (or boiling water) and watch it spin. The convection (upward motion) of hot gases provides the energy of motion.

7. Measure the temperature of the air above the heat source by holding the thermometer at least 25 cm above the boiling water.

This step will prove to you that gases above a flame or boiling water are warmer than the surrounding air.

© 1991 by The Center for Applied Research in Education.

Questions:

1. Do you think that this simple device can be turned into something with a practical use? Suggest uses for this principle.

   _____

   _____

   _____

   _____

   _____

2. Where in the room do you expect to find the highest temperature? Why?

   _____

   _____

   _____

3. What temperature did you measure in the convection air current above the candle?____

4. Is this simple engine a proof that heat is a form of energy? Is wind a form of energy? Find out more about wind and tornadoes and explain them.

   _____

   _____

   _____

   _____

   _____

   _____

   _____

5.5 ELECTROMAGNETIC SPECTRUM, RADIATION, AND INFRARED RAYS

Purpose: The purpose of this activity is to become familiar with the electromagnetic spectrum and selected forms of invisible energy.

Information: Energy travels in waves. Some waves, like light, are visible; other waves are invisible. Examples of invisible waves are: X-rays, heat (infrared), ultraviolet, radio or TV waves, radar, and sound. The entire system of all waves is the *electromagnetic spectrum*. Spectrum means band. You observe waves when you throw a rock in still water. The waves move (radiate) out and away. Energy works in the same way.

Radiation simply means the sending out of energy waves. A light bulb radiates light. A radio station radiates radio waves. A boiling pot of water radiates heat. An X-ray source radiates X-rays. A car radiator radiates out the car's heat energy.

Our sun provides many forms of energy besides light, which we can see and heat, which we can feel. These energy forms are: cosmic rays, X-rays, ultraviolet rays, infrared rays (heat), short and long radio waves. There is empty space between the earth and the sun. Since convection cannot take place without a fluid (gas or liquid), the sun's energy must travel in waves. Heat waves are invisible and are called *infrared*.

Equipment: Two aluminum cans (one painted black), two thermometers, graduated cylinder, tap water

Procedure 1:

1. Place 50 ml of tap water into each can and place them in the sun. Record the starting temperatures of water as the 0 minutes entry.

2. Record the temperatures of water about every five minutes until the end of the class period.

3. Predict which can will heat faster. If the sun is not available, the teacher will do the experiment and the entire class will share the data. He/she will place between the two cans a light bulb as a source of heat.

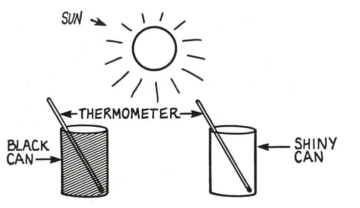

© 1991 by The Center for Applied Research in Education.

Sample Data Setup:

shiny can	time min.	0	5	10	15	20	25	30	GAIN
	temperature								
black can	time min.	0	5	10	15	20	25	30	GAIN
	temperature								

Calculations: To calculate the temperature gain of water, subtract the beginning temperature from the final one.

Questions:

1. Which can did you predict would warm up faster? How accurate was your prediction?

_____

_____

2. How warm did the waters become? How many degrees gain did each have?

_____

_____

3. What do these gains in temperature mean?

_____

_____

_____

Procedure 2:

1. Graph the values of the waters in the cans.
2. Place both curves on the same graph.
3. Draw each curve with a different color.
4. Label one "shiny can" and the other "black can."

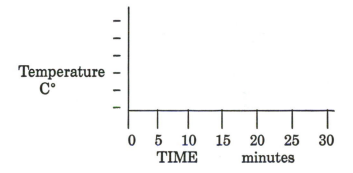

Questions:

1. Which color roof will be best to reflect the heat of the sun? What is your evidence? Explain.

   _____

   _____

   _____

2. Which color roof would absorb the most energy? What is your evidence? Explain.

   _____

   _____

Name _____ Date _____ Period _____

5.6 CALORIES IN FOOD

(Partner activity)

Purpose: The purpose of this activity is to measure food calories in one specific food item.

Information: Everyone has wondered at times how we know the number of calories in food. This information is printed on nearly all food products sold in the United States. Scientists have agreed that a calorie is the amount of heat which will heat 1 ml or 1 gram of water 1° Celsius. If you heat 1 ml of water 20 degrees higher than the temperature at the start, you have used 20 calories of heat energy. One food calorie is the same as 1000 science calories:

1 kilocalorie (science) = 1 food calorie (nutrition)

You will measure the calories of a food substance by building and using a simple calorimeter, a device to measure calories.

Equipment: Three wood dowels about 15 cm (6 in.) or pencils, candle, masking tape, thermometer, graduated cylinder, aluminum can, T-pin or needle and clothes pin, peanuts or popcorn, two shallow pans, one jar with water

Procedure:

1. Look at the illustration.
2. Assemble a calorimeter by taping three dowels to the can.
3. Place the assembly in one shallow pan.
4. Pour 50 ml. of water into the can.
5. Measure its temperature. This is the starting temperature.
6. Take a half of a peanut or a piece of popcorn and stick it with the needle. Turn the needle several times while holding the peanut.
7. Hold the needle with a clothes pin.
8. Light a candle; place it in the other pan and use it to ignite the peanut. Attach the candle to the pan with a couple of drops of wax; keep both pans together.
9. Place the peanut under the can to heat the water. You may want to turn the peanut over during its combustion. When the peanut fire ends, read the water temperature at once. All this time you have the jar of water handy, as a fire precaution.

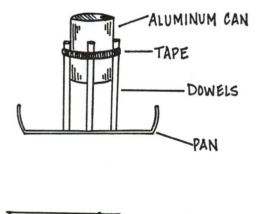

ALUMINUM CAN

TAPE

DOWELS

PAN

CLOTHES PIN WITH NEEDLE

PEANUT

10. Repeat this procedure three times and average your answers. You may use the same water: measure and record its temperature every time you start the procedure.

Sample Data Setup:

FOOD TESTED _____

Time	Starting Water Temperature C°	Ending Water Temperature C°	Temperature Change C° (A)	Volume of water ml (B)	Calories (A × B)
1					
2					
3					
				AVERAGE →	

Calculations:

1. Compute the difference in temperatures by subtracting the starting temperature from the ending one.
2. Calculate the calories by multiplying the difference in temperature by the volume of water.
3. If you need food calories, divide science calories by 1000.

Calories = milliliters of water × average change in temperature

Example: 100 ml of water changed 17°C. Calories = 100 × 17 = 1700 calories. 1700 divided by 1000 = 1.7 kcal. = 1.7 food calories.

Questions:

1. What is the number of calories in half of a peanut?_____

2. What is the number of calories in one whole peanut?_____

3. What is the number of calories in 25 peanuts?_____

4. What is the number of calories in 100 peanuts?_____

5. What is the class average of food calories for one peanut?_____

6. Assume that you burn a cookie (what a waste!) and find that it heated 1000 ml of water from 15°C to 70°C. How many calories are in the cookie? How many food calories?

_____

7. A. You have a *tablespoon* inside a *pot* full of boiling water. Which is hotter? Explain.

_____

_____

7. B. Which of these two objects has more stored heat energy?

_____

_____

_____

WAVES AND SOUND

TEACHER'S SECTION

In this section the following are touched upon:

1. The basics of wave shapes
2. Amplitude
3. Frequency
4. Wavelength
5. Speed
6. Sound

7. Resonance
8. Sound amplification
9. Dampening of sound
10. Ultrasonics
11. The Doppler effect

6.1 THE BASICS OF WAVES AND AMPLITUDE

The study of sound introduces waves. The basics of waves are easier to demonstrate with activities on sound. Traditionally these concepts are introduced with energy.

ACTIVITIES

Equipment: Slinky™, (optional) cathode ray oscilloscope (CRO), radio, phonograph, sine-square wave generator, test leads

1. Stretch a Slinky™ across the classroom. Have a student hold it. Shake it up and down to generate a sine wave.
2. Place one end of it next to a student's ear and pluck it. It will sound like a synthesizer.
3. If you have an oscilloscope, plug in a sine or square wave generator. For effect, hook up the generator to a speaker to show sound generation. Hook up the speaker leads of a radio to the scope to show audio waves. If you have the hook-up, show a wave and then show its amplitude increase as you increase the volume. Point out that waves have only five basic descriptors (attributes): *amplitude* (height), *wavelength* (width), *frequency* (how many per second), *speed* of travel, and *names* (if available).

Answers: See book for wave shapes.

Vocabulary: Wave, wave train, amplitude, sine wave

6.2 WAVELENGTH AND FREQUENCY

Make a sound by tapping a bottle, hit a ruler on a table, ring a bell, etc. Verify that students have an accurate understanding of wavelength. Wavelength is measured between the identical points of two consecutive waves, that is, at the crests, troughs, or reference line. Some books use the concept of period, the time for one wave to go by. Let the period $= T$, then frequency $F = 1/T$.

Answers:

1. See book for sample.
2. See book for wavelength diagram.
3. Speed = Frequency $\times$ Wavelength = 10 Hz $\times$ 20 cm = 200 cm/sec

Vocabulary: wavelength, frequency

6.3 SOUND AND HEARING: RESONANCE

1. Use tuning forks to make sound.
2. There may be some water splashing around. Place the tip of a vibrating tuning fork into a glass of water. The water stops the fork called *dampening.*

     Shock absorbers on the car dampen the mechanical energy of car vibrations, smoothing the ride.
3. Move about the classroom and touch your vibrating tuning fork to student desks. Ask students to place their ear on the desk surface. One vibrating object moves another. The desk will appear to make the sound louder. This is sound amplification.

Note: You use resonance to tune in radio and TV stations. When you tune in one specific station, you turn dials or push buttons. When the internal mechanism reaches the same frequency as the sending station, the sound and picture are received. Can you imagine the confusion if you were to tune in all the stations at once?

Answers:

1. A. vocal chords
   B. reed
   C. reed

   D. blades
   E. voice chords
   F. strings

Vocabulary: hearing range, resonance

6.4 STRING TELEPHONE

The voice vibrates the bottom of the tin can. The vibrations pass through the string to the other can. If you have a vacuum jar, demonstrate an electric bell's ringing while evacuating the air. When there is no air to transmit the vibrations, there is no sound.

6.5 SOUNDWAVE FIRE EXTINGUISHER

Answer:

2. The vibrating drum makes sound pressure waves, while the small point on the cone concentrates them to extinguish the flame.

Vocabulary: Extinguish

6.6 SOUND AMPLIFICATION

Mechanical amplifiers are large surfaces vibrating with the energy given by a smaller moving object. Think of a megaphone. Touch tuning fork to student desk.

Answers:

1. The larger vibrating surface acts as a mechanical amplifier.
2. Electronic amplifiers.
3. Similar to a synthesizer or reverberation unit.

Vocabulary: Amplification, supersonic, decibel, acoustics

6.7 UNDERWATER GUITAR

Build a simple underwater guitar by making a small wood frame about 12 inches long and 6 inches wide. Screw eye hooks into both ends and get new or used guitar strings. Stretch the strings between the eye hooks, twisting the hooks to increase string tension.

The water molecules absorb the energy of sound by dampening the motion of the strings. The shock absorber of a car dampens the vibrations caused by the road. You can stop the ringing of a bell by touching it.

The sound of a vibrating string changes with the following variables: thickness of string, length of string, and tension on the string. These laws of vibrating strings have been used and clearly understood since ancient times. If a string vibrates at 300 Hz and another one at 600 Hz, the 600 Hz is one *octave* higher. It vibrates at twice as many hertz as the basic note.

Answer:

3. The energy of the vibrating string was absorbed by the air and the water. Their absorption of energy makes them into dampeners.

Vocabulary: Sound, dampener

6.8 THE DOPPLER EFFECT

Activity: Use a simple flute, or a cheap whistle without a ball inside. This instrument must be able to produce a consistent tone (frequency of sound).

Procedure:

1. Ask students to close their eyes and to listen.
2. While blowing the instrument, walk briskly from one end of your classroom to the other.
3. Ask students to note the change in sound pitch. Many will report a sound similar to a train moving.

Answers:

2. A motorcycle going by or any other moving object
3. Yes, because there is a change in the pitch of the sound, higher when moving toward you and lower when moving away from you.

6.1 THE BASICS OF WAVES AND AMPLITUDE

Purpose: The purpose of this investigation is to learn about wavelength and resonance.

Information: Energy travels in waves. One can think of waves as vibrations. The key factor which makes one type of energy different from another is its *wavelength*. A *wave train* is a repeating set of regular waves. A wave train is the imaginary trace made by a piece of chewing gum stuck to the rim of a moving bicycle wheel. Let's plot its motion:

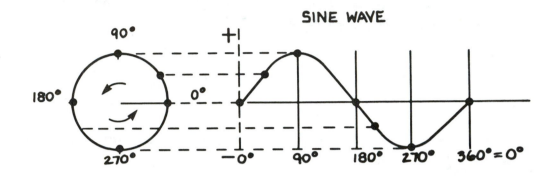

Look at the diagram above. At 0° the gum is on the right of the wheel and at 0 height. While going through a quarter turn, it rises to the highest point. At 180° it is back down again at the same height as it started. Next it goes below the axle of the wheel until it reaches the lowest point, having turned 270°. Now it turns and comes back to the same height where it started, having completed 360°. If this continues, one will have another wave. The wave just generated is a sine wave. Its height above or below the reference graph line is the *amplitude* (height), and it represents the energy of a wave. A wave with a greater amplitude has more energy than a wave with less amplitude. When you turn up the volume on a radio, you make the sound louder by increasing the amplitude (height) of electrical waves going to the speaker.

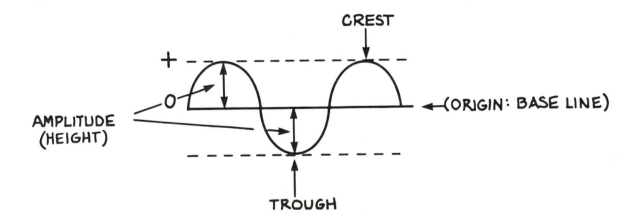

© 1991 by The Center for Applied Research in Education.

There are different names for differently shaped waves:

SINE WAVE

SQUARE WAVE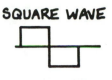

SAW-TOOTH WAVE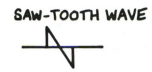

Activity:

1. Draw a sine wave.
2. Draw a square wave.
3. Drop a marble into a fishtank. Observe the waves on the surface.

6.2 WAVELENGTH AND FREQUENCY

(You need a partner.)

Information: The electromagnetic spectrum of energy has many waves, from extremely short ones to super long ones. Here is a partial listing: cosmic rays, X-rays, ultraviolet rays, visible light, infrared rays, radio waves, TV waves, microwaves (some radio, some TV, radar), supersonic sound, sound, and subsonic sound. The key difference between these different types of waves are their wavelengths.

Wavelength is the distance between the same points on two consecutive waves. The Greek letter lambda (λ) is used in science for wavelength. Think of it as a wave *width*.

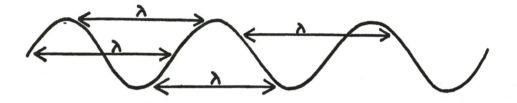

Another word associated with waves is *frequency*. It means how many waves per second a source generates. Its unit is the hertz (Hz). Its name honors Gustav Hertz for his pioneering work with waves. Radio and TV stations operate on very broad bands of frequencies. The Federal Communication Commission (FCC) assigns each station a specific frequency. Imagine that an FM radio station is at 98 on the dial. This means that it broadcasts 98,000,000 waves per second. It is written 98 MHz. *M* means Mega and stands for million in the metric system.

Look at the illustration. Imagine a box 1 second wide. If you increase the number of waves in this time box, the waves must have a shorter wavelength to fit. Wavelength is inversely proportional to the frequency. This means that as one increases wavelength, the frequency decreases.

To calculate the speed of a wave, use Speed = Frequency × Wavelength.

If you play with a rope which has 4 waves per second and each is 50 cm long then:

FREQUENCY

1 WAVE PER SECOND

2 WAVES PER SECOND

3 WAVES PER SECOND

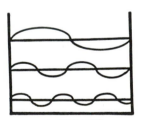

1 SECOND

Speed = Frequency × Wavelength
Speed = 4 Hz × 50 cm
Speed = 200 cm/sec

The answer means that the crest of the wave moves 200 cm/sec.

Activity:

1. Draw a wave train of sine waves.

2. Draw four wavelengths starting at different places on a wave.

3. Calculate the speed of a wave of 10 Hz, 20 cm long.

6.3 SOUND AND HEARING: RESONANCE

(Partner activity)

Purpose: The purpose of this activity is to understand sound.

Information: Sounds start with vibrating objects. These vibrations transmit through air to your ears. Tiny fibers in your inner ear sweep around and change the vibrations into electrical pulses. Your brain decodes these as sound.

The human ear receives vibrations from about 50 Hz to 16,000 Hz (vibrations per second). This is the human hearing range. Sounds above this range are supersonic sounds. Dogs and other animals can hear supersonic sound. Your dog probably howls long before you hear a fire siren, because he can hear the ultrasonic sounds of the siren. These ultrasonic sounds hurt his ears. People use ultrasonic (silent) whistles to call dogs. High frequency (ultrasonic) sound waves are used to shake the dirt off jewelry. Dentists use ultrasonic drills to reduce pain from drill sound vibrations.

Equipment: Two identical bottles, string, two spoons or forks, Slinky™

Procedure 1:

1. Take about 30 cm to 50 cm of string and tie it to a spoon or a fork.
2. Tie a large loop at the other end of the string.
3. Tie the loop on one of your ears.
4. Bend forward, so the silverware hangs from your ear, while your partner hits the spoon with another spoon or fork. You will hear chimes, like bells. The string is transmitting the sound vibrations.

Procedure 2:

1. Take one bottle and place its opening near your ear.
2. Have your partner from a couple of feet away blow across the other bottle to make a sound. Every time he does this, you will hear your bottle sound off. This phenomenon is *resonance*. When two objects can vibrate at the same number of vibrations per second, if one is vibrating, it sends out sound waves which vibrate the other one.

Questions:

1. Tell what is vibrating for the following sounds:
   a. human voice_____

   b. clarinet_____

   c. train whistle_____

   d. fan_____

   e. dog barking_____

   f. guitar_____

2. Name ten more sounds and tell what is vibrating.

   a. _____ _____   f. _____ _____

   b. _____ _____   g. _____ _____

   c. _____ _____   h. _____ _____

   d. _____ _____   i. _____ _____

   e. _____ _____   j. _____ _____

6.4 STRING TELEPHONE

Purpose: The purpose of this investigation is to demonstrate the action of sound waves.

Information: Solids, liquids, and gases transmit vibrations known as sound. Vacuum does not transmit sound. If someone wanted to shout at you on the surface of the moon, you could not hear her, even if she were standing right next to you. On the moon there is no atmosphere to transmit sound waves.

Equipment: Two tin cans of same size, or two cardboard oatmeal containers of same shape, with one end removed; two toothpicks and about 8 m (25 ft.) of string, one nail, hammer

Procedure:

1. Punch one small hole in center of can bottom. (Use a piece of wood below tin can).
2. Pass one end of the string through the hole in the bottom of can or box.
3. Tie a knot around a quarter of a toothpick (to prevent the string from coming out).
4. Repeat for other can and other end of string.
5. Stretch the string in a straight line. Now you can talk to other students through this tin can telephone.

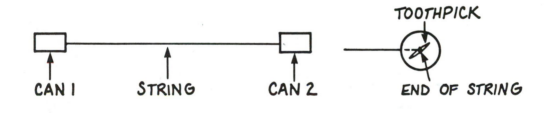

6.5 SOUND WAVE FIRE EXTINGUISHER

Home Project or Class Demonstration:

Caution: Adult supervision needed. Follow all safety guidelines.

Information: Sound waves provide pressure force (compression). This pressure and the lack of it between waves, moves our eardrums and we hear sounds. In this activity sound pressure controls the flame of a candle.

Equipment: Mailing tube (or cardboard tubing) about 3 in. to 4 in. (7 cm to 10 cm), rubber balloon, rubber band, sheet of paper, tape, candle

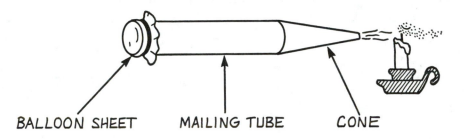

BALLOON SHEET   MAILING TUBE   CONE

Procedure:

1. Tie a single layer of rubber balloon material over one end of the mailing tube with a strong rubber band or string.
2. Make a paper cone and fasten it to the other end of the mailing tube.
3. Allow a hole about 1/4 in. or 0.7 cm on the small end of the cone. You have actually made a miniature model drum, and all sound waves must come out of the small opening.
4. Light a candle and place the tube over a book or other suitable support. Tape it in place so it will not roll.
5. Place the paper cone near the flame, as close as you can without burning the paper, then:
   a. Clap your hands near the rubber drum and observe the flame.
   b. Whistle near the drum and see the candle flame flicker and flutter.
   c. Play a small radio with its speaker near the drum.
   d. Flick your finger at the drum surface.

© 1991 by The Center for Applied Research in Education.

Questions:

1. What did you see?

_____

_____

_____

2. Explain how you extinguish the flame.

_____

_____

_____

6.6 SOUND AMPLIFICATION

(You need a partner.)

Purpose: The purpose of this investigation is to learn about sound, its amplification and the risks involved in this process.

Information: Many sounds are too faint to be heard. They must be made louder, a process called *amplification.* You can make a sound louder by either mechanical or electronic means.

Equipment: Tuning fork, a phonograph with an old record, needles, 2-by-4-inch cards, pins, paper

Procedure:

1. Hit a tuning fork on your shoe heel and touch the handle to a table. This will prove mechanical amplification. The table now vibrates in tune with the fork, producing the same number of hertzes. Since the table is a much larger vibrating surface, the sound is also louder.

2. Hold a needle in the groove of a moving phonograph record and listen carefully. You should hear a faint sound.

3. Pass the needle through a small card and repeat. The sound should be louder.

4. Make a paper cone, fold the small end over, and place the needle through the folded paper.

5. Hold the needle in the record grove and you have just duplicated how early record players amplified sound. This is the principle of the megaphone.

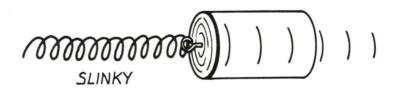

SLINKY

6. Take a Slinky™ and tie it to a cup hook fastened to the center of the bottom of an empty coffee can. Place a small piece of wood on the inside, so the hook will not come loose. Stretch the Slinky™ and tap it with a pencil and other objects. The can acts as an acoustic amplifier. You will hear some interesting sounds.

Information: Sound travels in air at 1087.1 feet per second or 33.136 meters per second at 0°C in dry air, at sea level. The speed of sound increases 2 feet for every increase of 1°C. At 25°C (75°F) the speed of sound is 1087.1 + (2 × 25) = 1137.1 ft./sec. In rough terms, the speed of sound is around 760 miles per hour at sea level.

When objects fly faster than the speed of sound, during a supersonic flight, they leave behind their own sound. Examples: the French-British Concorde, military jets. Supersonic flights generate high sound pressure waves which we call *sonic booms*. Glass objects may shatter from these high pressures. This is why supersonic flights are allowed only over high seas and other unpopulated areas.

Decibels measure sound pressure (loudness). A change of three decibels means either the doubling or halving of sound pressure. Decibels are logarithmic units, that is, they are nonlinear (proportional). If a person hears loud sounds, the ear drum tries to protect itself by growing thicker. As a result of loud sounds, the ear drum becomes less sensitive to weaker and normal levels of sound. This is one way of becoming deaf.

With modern electronic amplification, it is common to hear loud sound levels. You need to become aware of this serious problem with loudness of sounds. Protect yourself by either wearing protective ear devices or by going farther away from the source of sound. Sound levels drop rapidly with distance from their source.

Questions:

1. Why was the sound louder when you added the card and the cone to the needle on the phonograph?

   _____

   _____

2. Phonographs use _____ in place of the card and cone.

3. Describe the sounds you heard with the Slinky™.

   _____

   _____

6.7 UNDERWATER GUITAR

(Partner activity)

Purpose: The purpose of this investigation is to learn about sound dampening.

Information: Sound is the vibration of objects, such as solids and liquids, which in turn vibrate the molecules of air or other materials around them. Sound radiates (moves away in all directions) from the source of vibrations, until its energy is used up or absorbed by a dampener. Pianos have extra pedals to move felt pads to touch the vibrating strings, absorbing (dampening) their energy. Porous surfaces absorb sound energy at a higher rate than hard surfaces. Compare the sound of a radio or stereo in a room with bare walls versus the sound in a room with plush carpets and heavy drapes. The former room is *live* while the latter is *dead.* One needs a much more powerful high fidelity amplifier for the second room. Shock absorbers dampen (reduce the energy of) the vibrations of a car.

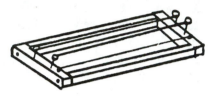

Procedure:

1. Listen to the sound of a vibrating string of a homemade underwater guitar in air and time its duration.

2. Place the guitar with vibrating string under water in a fish tank and listen. Have your partner place his or her ear on the fish tank. Record its duration.

Sample Data Setup:

Try #	Sound in air Time (seconds)	Sound in water Time (seconds)
1		
2		
3		

Average_____   Average_____

© 1991 by The Center for Applied Research in Education.

Questions:

1. How long did the strings vibrate in air? How long in water?

   _____

2. What is the difference in times? Why?

   _____

   _____

   _____

3. Where did the energy of the sound go?

   _____

   _____

   _____

6.8 THE DOPPLER EFFECT

Purpose: The purpose of this activity is to explore the Doppler effect.

Information: When sounds move, you observe that their sound changes in pitch. Think of a car passing while blowing a horn, or a train going by. As the sound moves toward you, it sounds higher; when it moves away it sounds lower. This is the *Doppler effect.* As a source of sound approaches an observer, the waves crowd together. The wavelength of the sound decreases, producing a higher pitch. As the moving sound source leaves an observer, the sound waves are further apart. The wavelength of the sound increases, producing a lower pitch. This phenomena is used for analysis of star spectra. It allows us to determine their motion, speed, and mass. Likewise, radar uses this principle to determine speed, etc.

Equipment: Electric buzzer, battery, coffee can, string, nail, tape recorder

Procedure:

1. Make several holes in the bottom of a coffee can.
2. Make one hole near lip of can.
3. Tie string into this hole.
4. Place buzzing noisemaker in can.

Caution: Make certain that no one is near the swinging can!

5. Fly can around in large circles.
6. Record sound of buzzer, just outside of its flying radius. By listening carefully to the sound of the buzzer, you will hear the change in its pitch.

Questions:

1. Describe the Doppler effect.

_____

_____

_____

2. Describe a couple of times when you might hear the Doppler effect.

_____

_____

_____

_____

_____

3. If you close your eyes, can the Doppler effect let you perceive motion? Explain.

_____

_____

_____

MAGNETISM

TEACHER'S SECTION

This section will cover the following concepts:

1. Magnetism
2. Electromagnets
3. Making electricity with magnets

In this chapter students will examine one of the invisible forces of the universe: magnetism. It has brought joy and amazement to people young and old throughout the centuries.

Lodestone (magnetite), a stone supposedly first found in ancient Greece, was one of the earliest discovered materials that had the special property of being able to attract iron. In modern times, stronger magnets and electromagnets were invented. Electromagnets behave as powerful magnets only while electric current is applied.

Iron materials are the only ones which can be picked up by a magnet. Steel is iron containing less than 1 percent of carbon.

Direct applications of electromagnets include door bells, switches in washing machines and appliances, electric relays, etc. A magnetic field surrounds wires carrying electric current. Using this and other simple principles, electric motors and generators were invented.

7.1 MAGNETS

Get small paper dishes from a restaurant supply house. You can purchase from science supply houses packages of 100 small, inexpensive (Alnico) magnets. Magnetic marbles, with their shells removed, are an inexpensive source of small but powerful magnets. You will need two bar magnets, one large strong magnet, and some iron filings; additionally, try to obtain a pair of wheel magnets. (Try Radio Shack.)

DEMONSTRATION: ATTRACTION AND REPULSION

1. Balance the bar magnet on a string loop and hang it near the front of the class for all to see.

2. Mark one end with a big red dot or bright red paint (if it comes unmarked). It will line up with the earth's magnetic lines. You just made a hanging compass.

3. Place a second bar magnet near the hanging one and watch them attract or push apart (repel). Mark the end of the hand-held magnet that repels the other red dot with a red mark. Review with class, after the boat races, the basic properties of magnetism which they have discovered.

Rule: Like poles repel, unlike poles attract.

DEMONSTRATION: "LEVITATION"

Place wheel magnets on a pencil and turn them so they repel each other. Tell class that this levitation (floating in air as if in defiance of gravity) is due to an invisible force. In science we observe many effects. From these, we must deduce the causes. This is circumstantial evidence.

For a more permanent display, drive a nail through a small board (3 in. by 3 in.). Place a drinking straw on the nail and the wheel magnets on the straw. A drop of glue will hold the straw on the nail.

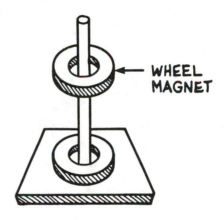

WHEEL MAGNET

DEMONSTRATION: MAN VS. MAGNET

If you have a powerful magnet, ask for a strong person who can remove a large magnet from a metal stool or other metal object. Allow everyone to try it and leave this magnet around for general class use. Keep the magnet away from magnetic tapes, computers, and diskettes!

DEMONSTRATION: MAGNETIC MARBLES

Purchase a couple of packages of magnetic marbles from a toy or science supply store. Purchase some regular marbles. Have students in small teams play with both types of marbles and compare what happens when two magnetic marbles meet. This activity is fun and most welcome to students. It should last just a few minutes.

Answers:

PROCEDURE 2

1. Probably yes

2. They either repel or attract. If the same poles come together, they repel; otherwise, they attract.

PROCEDURE 3

1. North and south poles
2. Same poles (− − or + +) repel; different poles (− + or + −) attráct.

Vocabulary: Magnet, magnetism, magnetic field, pole

7.2 MAKING A COMPASS

Ancient sailors became aware of compasses, but were never able to make really good ones. Their use made for erratic navigation at best. In the seventeenth century, the British Navy offered a large reward for an accurate clock which would work at sea. By knowing the exact time of the day and by using a sextant, sailors were able to pinpoint their positions at sea.

Answers:

1. The needle lines itself up along the north-south magnetic line.
2. The north end of the magnet is attractéd by the south pole's magnetic force, while the south end of the needle is attracted by the magnetic force of the north pole. Opposite magnetic forces attract each other.

7.3 ELECTROMAGNETS

Information: In these activities you show that (1) When electricity flows in a wire, there is a magnetic field around it; (2) The magnetic forces are radial and at 90° to the plane of the wire; and (3) A magnetic material can become itself a temporary magnet, while held by a magnet. (It transfers magnetic lines of force as if it were the magnet. Pick up a paper clip with a magnet and the clip will attract other clips. This is also true for electromagnets.) The last property is illustrated by tape recorder heads, which are electromagnets. After many cycles of gaussing (magnetizing) they retain enough residual magnetism to begin to erase and damage the magnetic tape. Also, the quality of playback will be poor. This is why you need to demagnetize the heads after every six or seven hours of operation.

When electricity flows in a wire loop, a magnetic field is produced and passed through the center nail. The nail becomes an electromagnet. In passing on the magnetic forces, the nail concentrates them, making them appear stronger. There are thousands of applications for electromagnets: to lift junk metal, on/off water switches in washing machines, in garage door openers, drivers for speaker cones, in burglar alarms, automobile starters, automobile horns, etc. Relays and solenoids are also electromagnets.

Vocabulary: Electromagnet, magnetic field

DEMONSTRATION: MAKE ELECTRICITY

Equipment: Loop of insulated bell wire (about 25 ft. to 30 ft.), bar magnet, projection galvanometer or transparent compass (from dime store) made into a galvanometer (wrap about 100 loops of #28 or finer wire along its N-S axis.)

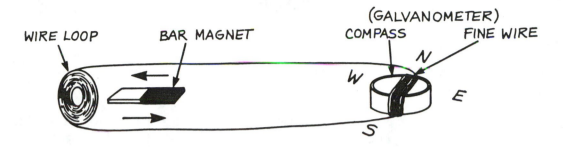

WIRE LOOP   BAR MAGNET   (GALVANOMETER)
COMPASS   FINE WIRE

PROCEDURE

1. Remove 3/4 in. of insulation from the ends of the bell wire and connect to the galvanometer.
2. Place galvanometer on overhead projector (focus the needle on your screen). A couple of students can describe their observations to the class, if there is no overhead projector.
3. Pass a bar magnet through the loop, stopping after each motion in, then out.

Information: By moving a magnetic field inside a wire loop, you have induced electricity in the wire (the opposite of the previous experiment). The galvanometer detects this electricity, since electricity moving in wires creates a magnetic field.

Answers:

PROCEDURE 1

1. Magnetic forces
2. In response to a magnetic force

PROCEDURE 3

1. The filings are attracted by the wire.
2. Magnetic force
3. The filings formed around the wire.

PROCEDURE 4

1. Yes
2. The more the wire turns, the stronger the magnetic field. Answers will vary.
3. In all types of electrical switches, like washing machine on/off water, horn relay in a car, etc.

Vocabulary: Electromagnet

7.1 MAGNETS

(Partner activity)

Purpose: The purpose of this investigation is to understand magnetism and some of its basic properties.

Information: Magnetism is a property of certain metals to attract each other. Magnets have two poles, one positive and one negative. Opposite magnetic poles attract; same poles repel each other. In a magnet, invisible lines of force, called magnetism, leave one pole to return at the other pole. Lines of magnetic force do not cross each other and are called a magnetic field. Magnetism is used in electric motors, speakers, microphones, and many other applications. Magnets are manufactured in many shapes such as bars, horseshoes, and circles. Lodestones are magnetic rocks. Their magnetism is weak compared to modern manufactured magnets.

Equipment: small paper cup, two magnets, water bowl, paper dots, bottle of red nail polish, fish tank, cellophane tape

Procedure 1:

1. Place a magnet on a sheet of paper.
2. Move the second magnet below the paper.
3. Try to move the upper magnet by moving the lower one.
4. Repeat using a thicker paper or a piece of cardboard.

Procedure 2:

1. Cut the upper two-thirds of a paper cup off (unless you have a small paper cup), to make it shallow.
2. Place a small magnet inside the cup and tape it to the bottom.
3. Float the magnetic boat in a small bowl of water (an empty cottage cheese container, for example).

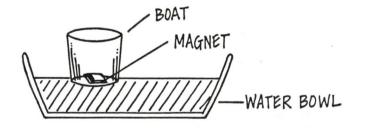

4. Play with the other magnet by placing it near the magnetic boat. Find out which two sides of both magnets attract each other.
5. Mark the end of the hand-held magnet with a label dot or red nail polish. Mark the opposite side of the boat magnet the same way.

Questions:

1. Were you able to move a magnet across a piece of cardboard? Explain why or why not.

_____

_____

2. What happens when you bring the second magnet near the magnetic boat? Why?

_____

_____

_____

Procedure 3:

1. Race your boat across a fish tank by using the second magnet. Time the event.
2. You and your partner compete for the best time.
3. All winners compete to discover the grand admiral, champion of the magnetic fleet.

Questions:

1. Name the poles of a magnet.

_____

2. Do same magnetic poles attract or repel? State the rule for how magnets work.

_____

_____

_____

7.2 MAKING A COMPASS

Purpose: The purpose of this investigation is to make a compass of the type used by ancient sailors.

Information: A compass is a magnetic needle which aligns itself along the lines of earth's magnetic lines. The magnetic north and south poles do not agree with the geographic north and south poles. When you use a magnetic compass, you must adjust for this error.

Equipment: small bowl (like cottage cheese container); water; disposable eye dropper, drinking straw, or cork; sewing needle; bar magnet

Procedure:

1. Magnetize needle by stroking it several times with magnet.
2. Press needle at 90° into cork or into a piece of soda straw about 5 cm long. If using a disposable dropper, cut ends off and use only the central portion.
3. Place cross-shaped assembly in a small dish with a few centimeters of water.

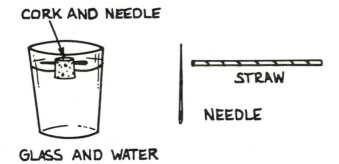

Questions:

1. What happened when the assembly was placed in water?

   _____

   _____

   _____

2. Why did the needle behave as it did?

   _____

   _____

   _____

Name _____ Date _____ Period _____

7.3 ELECTROMAGNETS

(Partner activity)

Purpose: The purpose of this investigation is to understand the principle of electromagnetism.

Information: Electric current flowing in a wire creates a magnetic field around it. This magnetic force is the evidence for electromagnetism.

Equipment: Wire loops (made of #28 wire or finer, about 100 loops 3 in. to 4 in. in diameter), 2 1/2 in. to 3 in. nail, paper clips, battery, lamp cord wire 2 ft. to 3 ft., small compasses (if not available go to Procedure 2), iron filings

Caution: Do not leave the loops connected to the battery for long time. The loops will get hot, and the battery will drain.

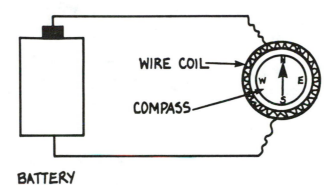

BATTERY

Procedure 1:

1. Place compass in middle of wire loop.
2. Connect loop to a battery by touching for an *instant*.
3. Repeat several times and observe the compass.

Questions:

1. If the needle of the compass moves, what is this evidence of?

_____

_____

2. When does the compass needle move?

_____

_____

Procedure 2:

1. Strip 1/2 in. of insulation at both ends of a 2 ft. to 3 ft. wire.
2. Place this wire on top of a piece of paper sprinkled with iron filings.
3. Connect wire to the battery for a short time.

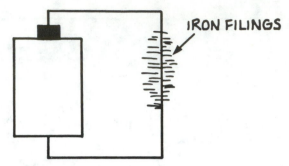

Procedure 3:

1. Pierce a small hole through the center of a piece of cardboard.
2. Insert a 2 ft. to 3 ft. insulated wire through this hole.
3. Sprinkle iron filings on the cardboard around the wire.
4. Connect this wire (with the ends bare) to the dry-cell for a short time.
5. Tap the cardboard gently.

Questions:

1. Explain in your own words what happened to the iron filings when you let current flow through the wire.

_____

_____

_____

2. What force do you think was present?

_____

_____

_____

3. What happened on the cardboard?

_____

_____

_____

Procedure 4:

1. Take wire loop and make one turn around nail.
2. Connect to battery and see if the head of the nail can lift any paper clips.
3. Repeat with two turns, then three, then four, and so on.

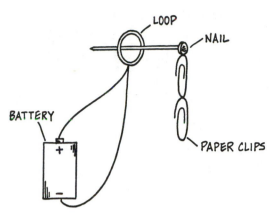

Sample Data Setup:

# of turns	# of clips lifted
1	
2	
3	
4	
5	
6	

Add more boxes if needed.

Questions:

1. Does the number of turns have any effect on the number of clips lifted? Explain.

_____

_____

2. If you wanted to pick up many paper clips, how many turns of wire would you use? Why?

_____

_____

3. Can you think of several uses for your electromagnet?

_____

_____

_____

8

ELECTRICITY

TEACHER'S SECTION

In this section the following concepts will be taught:

1. Static charges and their properties
2. Basic electric circuit
3. Series, parallel, and short circuits
4. A battery, the volt, amp, and ohm
5. A lemon battery
6. Conductors, insulators, and circuits

8.1 STATIC ELECTRICITY

You can present many of these demonstrations yourself and have students repeat them. Sometimes you will need to have groups of students to come and see these. If anyone wishes, have them come back for a second look and try. Before you show these, practice a bit. Remember that static experiments tend to fail on days with high humidity.

Explanation: Since the water molecule has both positive and negative charges, any outside charge would attract it. Opposite electrical charges (+ and −) attract. Like electrical charges (+ and +, or − and −) repel (push apart). The positive charges in water attract the comb. The balloons repel each other because they have the same charge (−). A student between two balloons discharges them because he or she is less charged with electrons.

Answers:

PROCEDURE 5

1. Yes, it seems to move closer to the plastic object.
2. They had the same charges. Like charges repel each other.
3. Repel, attract

Vocabulary: Static electricity

8.2 MORE STATIC ACTIVITIES

Your students will love this timely tip. To avoid getting static shocks on dry days, touch all door knobs first with a key which you are holding. This will discharge your body's static electricity.

An *electroscope* is a device which holds static charges, until cosmic rays (the universe showers us with them all the time) discharge them. This discharge became the evidence for the existence of cosmic rays. You may build a small electroscope or borrow one from another school. There are many manuals on how to build a simple one. Electroscopes are relatively inexpensive to buy.

A *transducer* changes energy from one form into another. A microphone is a transducer: it changes sound into electricity. A light bulb is a transducer: it changes electricity into heat and light. A fan is a transducer: it changes electricity into mechanical energy.

Demonstration:

1. Carefully have a student stand on four glass jars (insulators). Make sure that he/she is close to a water pipe or metal object.
2. Rub his/her clothing for at least a full minute with either a fur or a hot-water bottle.
3. Have someone touch his/her finger.
4. Repeat the charging and have him/her touch the water pipe.

Vocabulary: Static, electricity, transducer

8.3 BATTERIES AND BULBS

Here are the only possible working models:

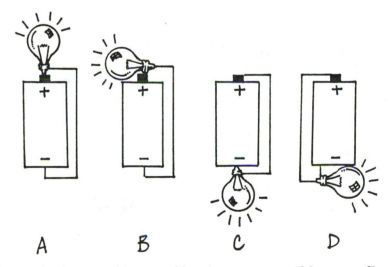

A   B   C   D

Be alert that only *four* working combinations are possible, regardless of how your student's work looks physically. The connections can be direct or through a wire. Sometimes students like to use two wires, etc.

1. Positive pole to base of bulb, negative pole to side.
2. Positive pole to side of bulb, negative pole to base.
3. Negative pole to base of bulb, positive pole to side.
4. Negative pole to side of bulb, positive pole to base.

You need to make clear that, by definition, all electric current in any circuit flows from the negative pole of the battery, goes through the circuit and then goes to the positive pole. Think of a sled ride down a mountain: the top is negative, the bottom is positive.

A bulb lights up because the filament inside opposes the flow of electrons. In this process the metal gets so hot it gives off light and heat. The bulb uses 99 percent of the electric energy for heat, 1 percent for light. Have students rub their hands: the resistance (friction) makes them warm. If electric current enters through the side of the bulb, it goes through the filament and exits at the center connection on the bottom. If electricity enters through the bottom, then it exits through the side. The bulb will work both ways. Bring to school a 25 watt *clear* light bulb for students to examine the inside.

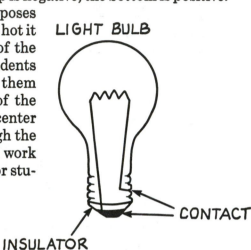

Answers: Typical findings by students are:

1. Several batteries make bulbs brighter.
2. Several bulbs on one battery make dimmer lights.
3. Batteries in opposite position (same poles facing each other) have their currents bucking (opposing) each other, and do not work.

Vocabulary: Battery, pole

8.4 SERIES, PARALLEL, AND SHORT CIRCUITS

You may wish to have pieces of wood (8 in. by 8 in.) with installed bulb sockets, switches, and battery holders. This is an excellent board for electrical investigations. If you use wire leads with alligator clips, you will avoid much aggravation with broken wires, tools, etc. Occasionally you may have to repair the clips. I do not specify the type of switch—many varieties are available. Touching two wires closes a circuit and is the simplest switch.

The series circuit has the disadvantage that if any part fails, the entire circuit is out. In parallel circuits, if one light fails the rest can still work.

Caution: Insist that during the short circuit activity, contacts are made only *momentarily.*

In a short circuit the current goes directly from the battery to the battery, bypassing all other parts of the circuit. Current always flows in circuits with the lowest resistance. The battery pushes out electrons without any limiting resistance like a bulb. It continues to provide power in increasing amounts. It gets hot and then fails. In real life situations, *wires get hot and fires start.* Fuses protect homes, automobiles, and most electrical devices. They work by melting, burning, or opening. Once the fuse is gone, electricity stops flowing in the line where the short circuits occur.

Answers:

1. In a series circuit, the electric current must go through each component (for instance, a light bulb). If any component fails, the circuit opens and does not work.

2. A parallel circuit allows each component to have its own hookup to the battery. If one component fails, the others continue to work.

3. When electrons can flow from the negative side of a power source to the positive side, without any load (components) in between such as bulbs, motors, etc., a large current is drawn. Either the wires or the power supply fail from the heat of the massive current. The circuit components, while using electricity, limit its amount in the circuit.

Vocabulary: Series circuit, parallel circuit, short circuit.

8.5 CHEMICAL BATTERY: MEASURING ELECTRICITY

It may be exciting to assign brief biographies of Volta, Ohm, Ampere, Galvani, Oersted, Watt, etc. These famous people were honored by having electrical units named after them.

Ohm's Law states: The voltage equals the current times the resistance. All discussions assume an electrical circuit.

Units: E = volts, I = amperes, R = ohms

$$E = I \times R \text{ or } I = E/R \text{ or } R = I/E$$

A toaster with 10 ohms resistance is plugged into 110V.

$$I = E/R = 110 \text{ volts}/10 \text{ ohms} = 11 \text{ amperes}$$

The mho (backwards spelling of ohm) is a unit of electrical conductivity.

Answers:

1. A. Carbon rod B. Zinc housing C. Chemical paste (electrolyte)

2. D.C. Direct current flows only in one direction in a circuit, from the negative to the positive pole of the battery.

3. A.C. means alternating current. Alternating current goes back and forth in a circuit. Typically, in the U.S.A., this happens 60 times per second.

4. An electrolyte is the chemical paste, liquid, jell, etc., which activates a battery.

5. The juice of the lemon, because it is acid.

6. The saliva, for it is acid.

Vocabulary: Electrolyte, electrode, pole, A.C., D.C., voltage, current

8.6 MAKING A LEMON BATTERY AND A GALVANOMETER

If you have a voltammeter, you do not need a galvanometer. Use the test circuits for current, usually labeled *mA* (milliamps). The leads must connect in *series* from one electrode (plate) to the other. In series means that electricity must go through the meter for the circuit to close. In the lemon battery, there is a scarcity of electrons in the copper strip. The zinc strip collects the excess of electrons. This charge difference causes electrons to flow through wires.

You may vary this activity by using a transparent compass, which is suitable for overhead projection.

There is a magnetic field around wires which carry current. With the compass aligned N–S, the magnetic field swings the needle in the E–W direction. The swing is proportional to the strength of the magnetic field, which depends on the amount of current. A weak current will swing the magnetic needle a little, a larger current more.

Activity For Home: You can make a neat alternative lemon battery by rolling a lemon, cutting two parallel slits and pressing into them a copper penny and a nickel. Leave coins barely above the surface. The coins can be very close but must not touch each other. By touching both coins with your tongue at the same time, you can feel a small electric shock. The lemon juice kills all bacteria on the coins.

Answers:

1. Parallel to increase current, series to increase voltage

2. Answers will vary. Most fruits have different pH values and amounts of juice.

3. A soft drink will work, because it is acid.

DEMONSTRATION: THE BAKING SODA BATTERY—CHARGING BATTERIES

Activity 1:

Information: When a battery delivers electric current, it is *discharging*. When a source of DC (direct current) electricity is connected to a battery, the battery is *charging*. The charging process restores the chemicals inside the battery to their former potential status. When recharged, the battery can be used again. (This discussion assumes use of a rechargeable battery.)

Equipment: A 300 ml beaker, two strips of lead about 5 cm (2 in.) by 15 cm (6 in.), baking soda, teaspoon, wires, flashlight bulb with socket or holder, 3 or 4 dry-cell batteries, electric train rectifier, or battery charger, sandpaper (#200 or finer).

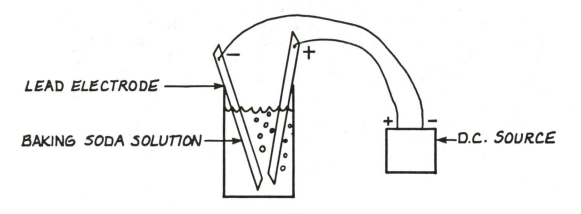

Procedure:

1. Fill beaker with warm water and saturate water with baking soda.
2. Punch a small hole in each lead plate, near one end.
3. Sandpaper the lead electrodes.
4. Place them in the beaker on the opposite sides.
5. Bend them out and down.
6. Connect to galvanometer or any voltammeter to show that there is no electricity available.
7. Connect this simple battery to a charged battery or a DC charger (HO train box, regular trickle car battery charger, or any DC source over 5 volts).

Note: There will be some bubbles appearing at both lead strips. The strip connected to the positive side of the batteries or the DC source will turn brown (lead oxide).

Activity 2: Connect the charged battery to voltammeter to show that battery is functional. Since it provides a very small current (quantity of electricity), it will not be able to light a bulb.

Questions:

1. What did you see when the battery was charging?
2. Did either electrode change color?

Vocabulary: Battery, galvanometer

8.7 CONDUCTORS, INSULATORS AND CIRCUITS

Note: The purpose of these activities is to introduce the idea that a modest assortment of parts can be used as real test equipment. Ideally, you want students to use two long leads as the switch. However, let them play until they arrive at this solution. The problem will demand it. They can use tape or their partner to hold the wires attached to the battery and bulb.

Note: Insulators can lose their insulating property and conduct electricity, if a sufficiently large electrical pressure (voltage) is applied. One inch (2.54 cm) of air will break down at about 30 kV (kilovolts). The electric pressure point at which an insulator breaks down and becomes a conductor is the *dielectric breakdown.*

Answers:

1. Most items will be made of metal.
2. They conduct electricity.
3. Cotton, paper, some plastics, wood, leather, rubber, etc.
4. Air is an insulator. If it were not, we would get shocked by the electricity from wires without insulation.
5. All metals are conductors. A few plastics are too.
6. They are usually nonmetals.

Vocabulary: Circuit, switch, conductor, insulator, semiconductor, transistor, integrated circuit, power supply

Name _____ Date _____ Period _____

8.1 STATIC ELECTRICITY

Purpose: The purpose of this activity is to demonstrate properties of static electricity.

Information: Most people think of static electricity as electric shocks they receive. Usually, this happens after walking on acrylic or nylon carpets or sliding around on plastic car seats. When objects or people rub other objects, they become charged with extra electrons. These are the *negative* charges from their orbit around the atoms. Electrons rub off easily from certain materials. When a charged object or person comes in contact with something *positive* (not charged up), the excess charges fly with a spark, like miniature lightning.

The problems of static electricity become severe during cold or dry weather. A statically charged person can totally destroy a computer diskette or electronic component by barely touching it. This is why computer equipment is usually grounded. It has leads to allow the excess electrons to flow to ground.

In homes and buildings, electric ground lines tie to cold water pipes, since these make a good contact with earth.

Demonstrations:

Equipment: Ping pong ball, string, 5 or 6 large balloons, fur or wool/nylon cloth, plastic comb, felt pen

Procedure 1:

1. Briskly rub a comb with a piece of fur, wool, or nylon material.
2. Bring the comb to a water faucet and place the comb near a fine stream of water (the finest you can get).
3. Notice the bending of the water stream. This is *electrostatic deflection*. It is easiest to see at the bottom, where the stream hits the sink.

Procedure 2:

1. Hang a balloon somewhere from a string at about head level.
2. Draw a smiling face on it.
3. When the face is dry, rub it with the fur.
4. As you come near it, it will try to kiss you.

Procedure 3:

1. Hang two balloons and charge them both by rubbing with fur.
2. They will stay apart because both have the same charges.
3. Place your head between the balloons. Both will kiss you.

185

Procedure 4:

1. Charge a balloon.
2. Place it on a surface like the wall.

Procedure 5:

1. Rub a *nonmetallic* object, like a plastic ruler. It will become charged.
2. Try to pick up little bits of paper, make hair stand up, or attract a ping pong ball.

Questions:

1. Did the water bend? Explain.

   _____

   _____

   _____

2. Why did the two balloons repel each other? Explain.

   _____

   _____

   _____

3. Complete the following sentences. *Like* static charges (+ and +) or (− and −) _____ each other. *Unlike* static charges (+ and −) _____ each other.

8.2 MORE STATIC ACTIVITIES

Purpose: The purpose of this activity is to continue to learn about static electricity.

Equipment: Newspaper page, aluminum pie tin, four identical Mason (glass) jars, bottle, hot water, piece of glass about 12 in. by 12 in. (30 cm), small fluorescent light tube, TV set

Procedure 1:

1. Cut out two newspaper strips, about 5 cm by 30 cm. Rub them with thumb and forefinger and watch them separate.

Procedure 2:

1. Take an aluminum dish (pie tin) and cover it with a sheet of glass.
2. Place inside the pan little cutout figurines of paper, no taller than the height of the dish.
3. Rub the glass with the fur and watch the figurines dance around.

Procedure 3:

1. Cut out a small paper frog. Place it beneath a glass sheet, propped up between two books.
2. Rub the glass with fur and watch the frog jump.

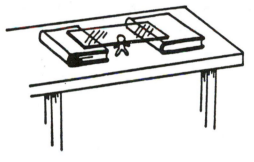

Procedure 4:

1. Place a newspaper page on the wall and stroke it flat with a pencil.
2. Lift a corner and watch it go back to the wall.

Procedure 5:

1. Cut out a tiny shape of a jet plane from aluminum foil.
2. Charge it with a charged plastic ruler.
3. Bring the charged ruler near the jet again. The plane will jump away.

8.3 BATTERIES AND BULBS

(Partner activity)

Purpose: The purpose of this investigation is to learn how an electric circuit works.

Information: Batteries are chemical devices which push electrons around circuits, acting as pressure pumps. A flow of electrons is an *electric current*. A tire pump pushes (pressurizes) the air inside tires, but it does not make the air it pushes. A battery pushes electrons around the circuit, but it does not make them. Batteries use chemicals to change chemical energy into pushing power.

The unit of electrical pressure energy is the *volt* (V). Once the chemicals in the battery are used up, the battery is dead. Some batteries like nickel-cadmium and car batteries (wet cells) can be recharged. Charging batteries reverses the chemical processes inside the battery, making chemical energy again available for use as an electrical pressure source.

Whenever the bulb lights up, you have a *closed circuit*. In it the electrons can leave the negative (−) pole of the battery and return to the positive (+) pole.

Equipment: One bulb, one battery, one wire 5 in. to 10 in. long

Procedure 1:

1. You will hook up a battery and a bulb in as many ways as you can.
2. Draw a picture of every different working model.

Caution: When making contacts, hold them only for a brief time. When the battery's positive pole connects directly to the negative pole, through any combination of wires, you have a *short circuit*. A short circuit will destroy the battery and make the wires hot. This can cause severe burns. Be careful to avoid short circuits. You will need the batteries a few more times.

Procedure 2:

1. Repeat activity using two bulbs, two batteries, and two wires. Share the equipment with another student.
2. Keep sketches of connections and results.

Questions:

1. Can you draw a picture of the bulb? What does it look like inside?

2. Why does the bulb light up?

_____

_____

_____

3. Why do you think bulbs light up only with certain hookups?

_____

_____

_____

8.4 SERIES, PARALLEL, AND SHORT CIRCUITS

Information: Electrical wiring uses two basic circuits: the series circuit and the parallel circuit.

Imagine that you are a tiny electron leaving the battery at its negative pole. In a *series circuit* you have only one path to go on your way to the positive end of the battery. In a *parallel circuit* you have two or more choices of how to go.

Equipment: Two miniature light bulb sockets with bulbs, wires or leads with alligator clips, battery, battery holder, switch

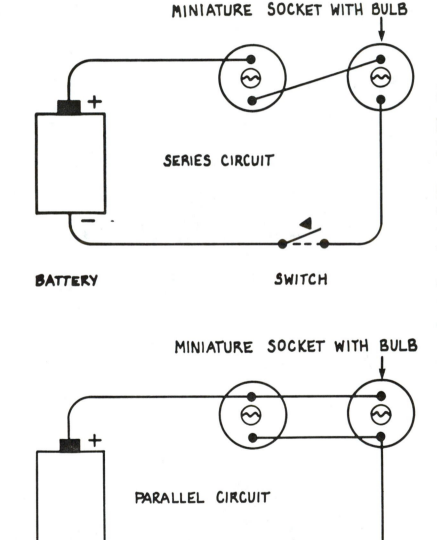

Procedure 1:

1. Hook up equipment as in illustration at right. This is a series circuit.
2. Close switch.

Procedure 2:

1. Hook up equipment as in the illustration at right. This is a parallel circuit.
2. Close switch.

Procedure 3:

1. Hook up equipment as in illustration at right. This is a normal series circuit.
2. Close switch. The light bulb will light up.
3. Connect one end of a wire (or lead with clips) to terminal *A*.
4. Connect only momentarily the other end of the lead to point *B* and observe. When you connect the wire from *A* to *B*, you have a short circuit.

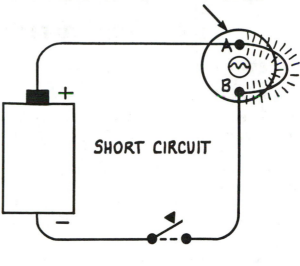

MINIATURE SOCKET WITH BULB

SHORT CIRCUIT

BATTERY   SWITCH

Questions:

1. Describe a series circuit.

_____

_____

_____

2. Describe a parallel circuit.

_____

_____

_____

3. Explain what happens when there is a short circuit.

_____

_____

_____

8.5 CHEMICAL BATTERY: MEASURING ELECTRICITY

Purpose: The purpose of this activity is to learn how a battery works and the basic units of electrical measurement: volt, amp, and ohm.

Information: In batteries, the *electrolyte* (a chemical) interacts with two or more different metal-carbon strips (electrodes). This causes a flow of electrons.

Electrons flow from areas of excess electrons to areas with fewer or no electrons. This difference in electrons is the *potential difference*. The greater the difference, the greater the push. Think of a slide: the steeper the slope, the greater the sliding force down hill. The pushing pressure of electrons is *voltage,* and the unit of its measurement is a volt (V). Most ordinary batteries are 1.5 V.; most car batteries are 12 V. The electricity in the United States is 120 volts *alternating current* (AC).

The moving of electrons in a conductor is electric *current.* The quantity of electric current going by, at any one point in a conductor, is *amperage* and the unit is the *ampere* (A). Materials which oppose the flow of electricity have *resistance*. The unit of resistance is the ohm (Ω).

Electricity from any battery source is *direct current* (DC). If a bulb is attached to a battery, the electric current *(I)* flows only from the negative pole to the positive pole. Alternating current continuously reverses its direction. One moment it flows from the positive to the negative pole, the next moment from the negative to the positive pole. In the United States we use electricity with 60 cycles, which means that it reverses itself 60 times in one second. Many countries use 50 cycles and different voltages. In those countries, one cannot use electrical appliances made for the U.S.A.

Information: It may come as a surprise to you, but dry cells are not really dry. The chemical paste (ammonium chloride + water + powdered carbon) acts as the electrolyte between the carbon rod and the outer zinc housing. When the battery discharges, hydrogen bubbles coat the carbon rod. This decreases the flow of electrons, because hydrogen acts as a resistor. Manganese dioxide, in the paste, reacts with and removes the hydrogen, making extremely minute amounts of water. When the cell becomes dead, its internal paste is dry. Manufacturers cover the cells with a steel housing, to keep moisture in and to prevent the chemicals from spilling out.

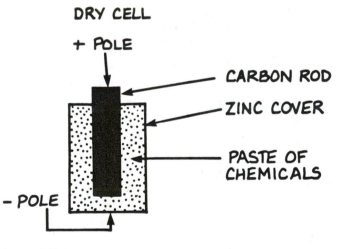

DRY CELL

+ POLE

CARBON ROD

ZINC COVER

PASTE OF CHEMICALS

– POLE

Another type of battery is a wet cell. A good example of a wet cell is a car battery. The liquid electrolyte in a car battery is diluted sulfuric acid.

You may recall receiving a mild electric shock by touching a fork or spoon to one of your teeth. Your filings (silver) and the fork (iron-steel), together with your saliva as the electrolyte, made a battery. The same happens if you leave some aluminum foil on your baked potato and take a bite. Ouch!

Questions:

1. Name the three basic parts of a zinc-carbon battery.

   A._____ B._____ C._____

2. What kind of current does a battery provide? _____. Explain it.

   _____

   _____

   _____

3. What is meant by *AC?*_____. Explain it.

   _____

   _____

   _____

4. Explain what an electrolyte is.

   _____

   _____

5. If you make a battery with a lemon and two coins of different metals, what is the

   electrolyte?_____

6. When you touch aluminum foil with your teeth, you get a shock. What is the electrolyte?

   _____

8.6 MAKING A LEMON BATTERY AND A GALVANOMETER

Purpose: The purpose of this activity is to learn the basics of batteries by building a lemon battery.

Information: A battery is a chemical device which pushes electrons in a closed electrical circuit. This action lasts as long as the battery has active chemicals to provide the pressure. A lemon contains acid juices, which can be used as the electrolyte to make a battery. An *electrolyte* is a chemical that activates a battery. Electrolytes are either acids or bases (alkali). Usually, they are a liquid, a paste, a jelly, or a gas. Electrodes are the materials in contact with the electrolyte, and they become the battery's poles.

A galvanometer is a device which detects very weak electrical currents.

Equipment: Lemon, magnetic compass, small spool of fine wire (#26–#28 or finer), a strip of copper (or pennies), a strip of zinc 1/2 in. by 3 in.—maybe the outside metal of an old dry-cell battery or a piece of zinc-coated (galvanized) sheet metal, scissors, sandpaper (#200 or finer)

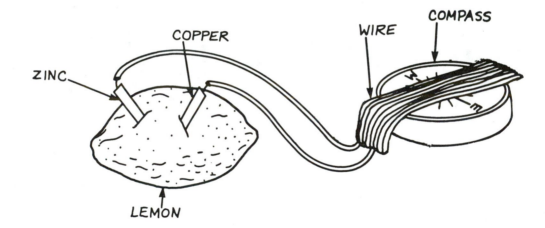

Procedure: To build a galvanometer:

1. Place about 100 turns of wire around the compass. Twist the free ends together a few times so they don't come apart.

2. Leave about 12 in. of wire.

3. Scrape off the insulation from the last 1/2 in. of the wires; use the sandpaper or a sharp blade.

© 1991 by The Center for Applied Research in Education.

Procedure: To make a lemon battery:

1. Roll a lemon on its side, pressing down with your hand, so the inside breaks up and the juices flow freely. The lemon contains citric acid, which is the electrolyte for the lemon battery.
2. Cut two half-inch slits in the lemon and insert the copper and zinc strips (electrodes). Make sure that the electrodes do not touch each other inside the lemon.
3. Connect the galvanometer to the electrodes. Make sure that the galvanometer is level and that the compass needle can move freely. The needle will move when there is a current flow. Open and close the circuit and observe the needle of the compass.

Questions:

1. Can you increase the current by using several lemons and connecting them? How would you connect them, series or parallel? Why? Explain.

_____

_____

_____

2. Try different fruits for batteries and tell which has the best electrolyte.

_____

_____

_____

3. Could you use a soft drink in place of a fruit? Explain.

_____

_____

8.7 CONDUCTORS, INSULATORS, AND CIRCUITS

(Partner activity)

Purpose: The purpose of this activity is to provide an understanding of insulators, conductors, switches, and circuits.

Information: A *circuit* is the path of the electrical flow. When electrons leave a power supply (a source of electric power, DC or AC) and return to it, that is a *closed circuit*. In an *open circuit* there is no electrical flow. If you touch any two objects (in a circuit with a battery and a bulb) and the bulb lights up, you have just made a *switch*. A switch is a device which opens and closes electrical circuits. *Conductors* are materials which allow the flow of electricity; insulators hamper the flow of electricity. *Semiconductors* are devices that behave like both conductors and insulators. A diode is such a device. It allows direct current (DC battery power) to flow in one direction only. Transistors and integrated circuits (ICs) are additional examples.

Equipment: Battery, bulb, 2 or 3 wires 6 in. (15 cm) long with stripped ends, adhesive tape, paper clips, ordinary string 3 cm to 5 cm (1 in.), some masking tape, paper plate or plastic bag (to hold everything)

Procedure 1:

1. Use the battery, bulb, and one wire to close the circuit. Try until the bulb lights up. Then make the light blink.
2. When you succeed, draw the picture on your data sheet and label it #1.

Questions:

1. Can you make the bulb light up and blink by assembling the three pieces in other ways? If you can, draw their pictures below and call them 2, 3, etc. Tell how they are different from your original #1.

_____

_____

_____

Configuration 1	Configuration 2	Configuration 3

Procedure 2:

1. Using your new blinker, (you may use two pieces of wire), connect a paper clip into the circuit.
2. Connect a piece of ordinary paper in the circuit.

Procedure 3:

1. Use your newly designed blinking test device to check if other items can open or close your circuit. Try your pencil, the eraser, the metal around the eraser, your pen, your desk, desk leg, the chair and other objects around the room.
2. Make two lists and separate all items tested into two groups:
   a. Those which close the circuit (conductors)
   b. Those which do not close the circuit (insulators).
   Test enough items to have at least twelve listings in each data group. If you are not too sure of what metal something is made of, just write "metal." One example of each is provided.

Sample Data Setup:

ITEMS THAT DO NOT CLOSE THE CIRCUIT
INSULATORS

	ITEMS	made out of
	PAPER	WOOD
1		
2		
3		
4		
5		
6		
7		
8		
9		
10		
11		
12		

ITEMS THAT CLOSE THE CIRCUIT
CONDUCTORS

	ITEMS	made out of
	COIN	METAL
1		
2		
3		
4		
5		
6		
7		
8		
9		
10		
11		
12		

Questions:

1. List a few materials that close the circuit.

   _____

   _____

2. Why do you think that materials that close a circuit are called conductors?

   _____

   _____

3. List a few materials that leave a circuit open.

   _____

4. Is air a conductor or insulator? Explain.

   _____

   _____

5. Did the conductors have anything in common? Explain.

   _____

   _____

6. Did the insulators have anything in common? Explain.

   _____

   _____

9

LIGHT

TEACHER'S SECTION

In this section students will have activities to prove properties of light. The key concepts demonstrated by the activities are:

1. Incidence and reflection of light
2. Refraction of light
3. Lenses and eyesight
4. Diffraction or scattering of light
5. Primary colors
6. Chromatography and capillary action

9.1 THE REFLECTION OF LIGHT

When you look at a light, you see the source. If you shine the light on an object, you see reflected light.

Equipment: *Mirrors*: You can use pocket mirrors and place them on a stand of clay or buy about eight of the least expensive mirror tiles and one glass cutter. Cut the tiles into six rectangles 4 in. by 6 in. each. Have the shop or a friend who has wood tools provide you with a piece of wood 1 in. by 2 in. by 6 in. It needs a 1/8-in. groove about 1/4 in. deep in the middle of the 2-in. side, together with matching pieces of 5 in. by 6 in. Masonite board (the type used for clipboards). Glue the boards into the slot at 90° to the base and, using white glue, fasten one mirror to each stand. You will have an inexpensive set of good laboratory mirrors. For safety's sake, cover the exposed edges of the mirror with masking tape.

Light: Use a flashlight or a slide, film, or overhead projector.

9.2 INCIDENCE AND REFLECTION

One of the crucial elements in this investigation is to help students to see that one vertical pin covers the view of the one behind it. You need to move the head along the desk and to be slightly above the level of the pin. Have students practice this with two pins at first. Give them some practice with protractors. If students are familiar with geometry and have compasses, have them draw a line perpendicular to *XY* at point *B*.

Vocabulary: Incidence, reflection, normal (angle)

9.3 REFLECTION AND REFRACTION OF LIGHT

Multiple images are possible, because reflected images are as far behind the mirrors as the real objects are in front of them. Images behind the mirrors reflect from one mirror to the other. Refraction or bending of light happens when light changes the media through which it travels. Each medium has a different density. (*See also* 3.7 Schlieren.)

Vocabulary: Refraction

9.4 LENSES AND EYESIGHT

Answers: 1. Convex 2. Concave 3. Convex

Vocabulary: Concave, convex, focus

9.5 COLORS AND RAINBOWS

I advise that you have individual student spectroscopes for at least 25 percent of your students. These are sturdy and can measure light in angstroms: visible light goes from about 4000 Å to 7000 Å (1 Å = 1/10,000,000 millimeter). You need also a few prisms and diffraction gratings. Get from any scientific supply house an 8-by-10-in. sheet of diffraction grating, then cut it into small 1-by-1-in. squares. If the grating is not available, take a microscope slide, stretch on it a piece of black electrical tape and cut a slit in the middle, all the way. The tape will separate slightly and provide you with a viewing slit. The image will not be as sharp as with professional tools, but a spectrum will appear.

Caution: Purchase low wattage bulbs. The small 7Watt night-light bulbs work well. If you use a larger bulb, use a variac or light dimmer to decrease the brightness of the bulb.

A neat activity is to change the brightness during the observation. Students will see a change in the spectrum, due to the change in the temperature of the filament.

Have students hold the slide to their eyes and look at the filament of a clear bulb from 3 in. to 5 inches away.

Demonstration: Place a mirror under water in a pan and shine light from a projector on it. The mirror will reflect a rainbow. Stir the water to show what happens. Let students explore this setup.

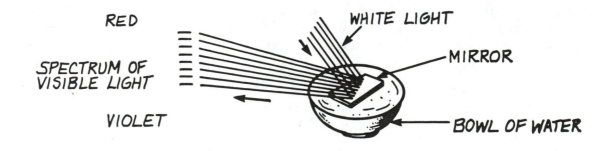

RED

SPECTRUM OF
VISIBLE LIGHT

VIOLET

WHITE LIGHT

MIRROR

BOWL OF WATER

Demonstration: If you have a black-light source, shine it on a TV screen and watch the screen glow in the dark. The phosphorus in the TV screen is a transducer (changer) of energy from one form into another. This is also an example of phosphorescence. Try your black light on clothes, rocks, and various plastic key chains.

Vocabulary: Spectroscope, spectrum, incandescent, fluorescent, phosphorescent

9.6 COLOR COMBINATION AND REFLECTION

Theatrical gels are a good source for color filters. Many theatrical supply houses send free preview samples. An alternative to flashlights is to use three (slide, film strip, or movie) projectors for a class demonstration.

DEMONSTRATION:

Equipment: Light cardboard, compass, plastic wrap, crayon, ruler, scissors, small motor or hand centrifuge or electric drill

Procedure:

1. Make a wheel of cardboard about 20 cm (8 in.) in diameter.
2. Section it off into 6 to 9 wedges.
3. Color the sections with alternating red, blue, and green.
4. Place the wheel (after cutting a small hole in center) on a record player set at 78 rpm. If you use a small electric motor, an electric drill or a hand centrifuge, make the color wheel only 3 in. in diameter. As the wheel turns rapidly, the colors should blend into one light creamy color. If any one color dominates, scrape some off.

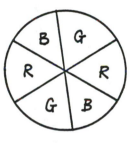

Homework Activity: Ask students to look at a color television set with a pair of binoculars. The binoculars need to be reversed, and the oculars must touch the television screen. While observing, they should turn off the color volume control, so that the image becomes monochrome. On some TV sets they may have to disable the automatic color switch to turn off the color manually. For comparison, if students have monochrome sets, have them look at the B/W screen in the same manner.

Explanation: Color television picture tubes have a large metal mask behind the glass screen. The screen surface is made up of thousands of triads of green, blue, and red phosphorus dots or stripes. As the electron beam strikes each triad, it provides a dot of composite color. To accomplish this, the electron beam changes its intensity and variation of red/blue/green strikes. As one reduces the color volume, to obtain a monochrome picture, the triads remain the same. On a monochrome screen, there are no triads. If a television set provides incorrect colors and fringing around the images, a technician is called to aim the electron beam to hit each specific triad accurately. This process is called *color convergence.*

Vocabulary: Absorb

9.7 CHROMATOGRAPHY AND CAPILLARY ACTION

The activities in this investigation may take more than one day. If you wish to blend colors for interesting samples, use the chart:

COLOR BLENDING CHART FOR FOOD COLORS

Number of drops required				
	GREEN	YELLOW	RED	BLUE
ORANGE		2	1	
PURPLE			3	1
TURQUOISE	1			3
CHARTREUSE	1	12		
TOAST	1	4	3	
VIOLET			1	2

Vocabulary: Component, absorbent, capillary, color

9.8 CAPILLARY ACTION

In Procedure 2 you use food coloring, since the width of a color stain is easier to see and measure than that of a clear drop of water. Give preference to dark colors.

You may wish to discuss other examples of capillary action (as in trees). Water rises due to:

1. Capillarity (small pores)
2. The cohesion of water molecules to each other
3. The strong adhesion of water to wood. Consider that many trees are taller than 100 ft. They have roots which are as deep as the trees are high. Water must rise to the top branches to bring minerals and nutrients.

Name _____ Date _____ Period _____

9.1 THE REFLECTION OF LIGHT

(Partner activity)

Purpose: The purpose of this investigation is to learn about a property of visible light: reflection.

Information: Light is a form of visible energy that travels in a straight line. This makes it impossible for us to see around the corner, unless we bend light using a mirror, fiber optics or Lucite™ (a plastic). You may recall using a mirror to reflect sunshine on objects and persons. Your eyes can see either the light source or light reflected from material objects.

Equipment: Mirror on stand, florist clay, pins, ruler, protractor, pencil

Procedure 1:

1. Take a page with writing on it and face it away from you.
2. Lift a mirror and look at the page through the mirror.

Question:

1. Why is the page backward?

_____

_____

Procedure 2: Use a light source to shine light onto the ceiling.

Question:

1. What kind of light do you see when you look at the ceiling?

_____

Procedure 3:

1. Insert a pin in a piece of clay.
2. Place the clay in front of the mirror, about 15 cm away.
3. Point your pencil behind the mirror until the pencil seems to touch or to be at the same distance as the image of the pin reflected in the mirror. At that point, the pencil is at the distance that the real object would be behind the mirror, if it were possible. The distance from the object to the mirror is the *real distance*. The distance from the front of the mirror to the image in the mirror is the *depth distance*.
4. Place the pin at several distances and record both the real and the depth distances. Graph these values.

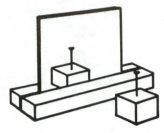

Sample Data Setup:

Real Distance cm	Depth Distance cm

Depth Distance cm

0   1   2   3   4   5
Real Distance cm

9.2 INCIDENCE AND REFLECTION

Purpose: The purpose of this investigation is to learn that visible light has equal angles of incidence and reflection.

Information: Good players of pool or billiards know that when balls hit a table edge, they bounce off at the same angle at which they arrived, unless they were given special twists. The same is true when you bounce a ball on the ground. The angle at which the object strikes the surface is the *angle of incidence.* The angle at which the object leaves the surface is the *angle of reflection.* Scientists have found that for visible light, the angle of incidence equals the angle of reflection.

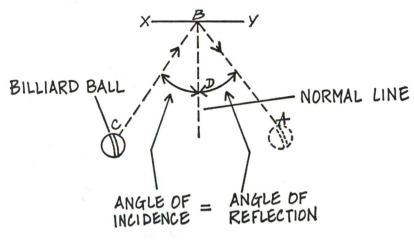

Information: A line which is perpendicular to a surface (at 90°) is a *normal* line. You measure from a reference normal line the angles of incidence and reflection.

Procedure:

1. Place a sheet of paper with the long side facing you.
2. Draw a line 5 cm (2 inches) from upper edge *horizontally across page.*
3. Label the line *XY.*
4. Select a point near the middle of line *XY* and label it *B.*
5. Draw a *dotted* line at 90° to line *XY,* through point *B.* Label one of its points D. See diagram for reference.
6. Draw a line from point *B* toward the right corner of the paper.
7. Mark a point on this line about 3 or 4 in. from the edge of the paper. Label it *A.*
8. Place the mirror along line *XY.* Make sure that *XY* lines up with the back edge of the mirror.
9. Take two pins, stick them on small pieces of clay, and place them on line *AB.* Make sure that the pins are vertical.

10. Lower your body to the level of the pins. Move your head left and right. Look at the pins toward the mirror, until the front pin covers the back pin and they look like one.

11. Place a pin to the left of line *BD* and move it around, until it looks as if it is lined up in the mirror and covers the other two pins. All pins should appear in a straight line. Keep your eyes near the level of the pins. After you line up the three pins, press the third pin down to mark the point. Label it *C*.

12. Draw a line from *B* going through point *C*.

13. Measure angles *ABD* and *DBC*. Repeat investigation until both measured values are equal.

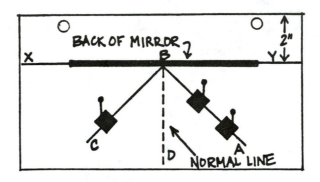

9.3 REFLECTION AND REFRACTION OF LIGHT

(Partner activity)

Purpose: The purpose of this investigation is to develop an understanding of reflection and refraction of light.

Information: *Reflection* is the return of light after it strikes a surface. Good reflectors include mirrors, most polished surfaces, and the still waters of lakes and ponds.

Equipment: Two mirrors, candle, protractor, paper clip, penny, glass

Procedure:

1. Place two mirrors at 90° to each other. You will see the original candle and three images.
2. Place the mirrors at 60° to each other. You will see six candles.
3. Try smaller angles and you will see more candles.
4. Place a candle between two parallel mirrors. You will see many candles.

MIRRORS AT 90°

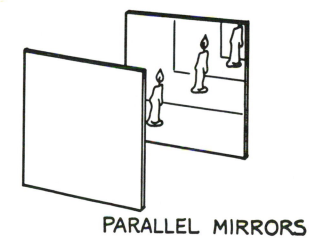

PARALLEL MIRRORS

Information: As light travels through different transparent materials, its rays bend. This is *refraction.*

Procedure:

1. Bend the end of a paper clip to make a tiny loop. Place in the loop one drop of water. This device inspired long ago the idea of a microscope. You have made now a device to enlarge objects too small to see with the unaided eye.

2. Look at several objects around you: a hair on your hand, your finger, your watch, the pencil tip, etc.

3. Place a pencil in a glass almost full of water. Look at the glass from several angles and see the pencil bend.

4. Fill a glass one-third full of water and drop a penny on the bottom. As you look at it from the side, you will see two pennies. Shake the water to find out which is the real penny. (The real penny will not shake.) You probably notice that the real penny appears larger than it actually is. This is because the water acts as a lens. Swimming pools appear shallower than they really are because the water magnifies their bottom.

Name _____ Date _____ Period _____

9.4 LENSES AND EYESIGHT

(Partner activity)

Purpose: The purpose of this investigation is to learn that different types of lenses can correct human eyesight.

Information: Plastic or glass lenses change the direction of light rays. Lenses are used in glasses, cameras, telescopes, etc. A *convex* lens is recognized by its thickness in the middle. A convex lens brings the rays of light together near the lens. As the lens becomes thicker, the focus (point where all the rays of light come together) moves closer to the lens. Convex lenses are used to magnify objects and can burn a hole in a piece of paper. Convex lenses are used in correction glasses for far-sighted people.

Far-sighted people, who cannot see clearly nearby, have the retina too close to the eye's lens. Before correction, images appear behind the retina and the people see objects blurred. The convex correction lens moves the focus point forward. Now images appears in the retina and not behind it. You made a convex lens out of a drop of water, remember?

Concave lenses are thinner in the middle and thicker on the edges. They scatter the rays of light. As the center becomes thinner, the scattering of light increases. Concave lenses are used for near-sighted people. Near-sighted people have the retina too far from the eye lens. The image forms ahead of the retina and they see images blurred. By scattering the light, the concave correction lens moves the image back to the retina.

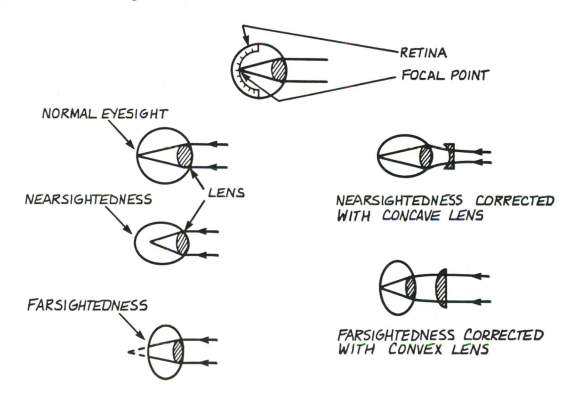

Equipment: Vial with cap or baby-food jar with lid, zip-lock bag, water

Procedure:

1. Fill vial with water, cap it, draw a line on a piece of paper, place vial over the line, and observe.
2. Place vial over written words.
3. Look through the vial at objects and room lights. Notice the scattering of light. You just made a convex lens.
4. Fill the zip-lock plastic bag with water, seal it well and use it as a convex lens to make additional observations.

Information: The human eye has a lens which converges light and focuses images on the retina. These images turn into electrical impulses. As the impulses reach the brain, they are interpreted as images.

Questions:

1. What kind of lens do humans and animals have?

_____

2. What kind of lens does a near-sighted person need?

_____

3. What kind of lens does a far-sighted person need?

_____

9.5 COLORS AND RAINBOWS

Purpose: The purpose of this investigation is to find out that light is several basic colors mixed together.

Information: In a rainbow, droplets of water act as prisms and scatter light, producing the colors of the rainbow. These colors are the basic parts of white light. Each color has its own wavelength, deep blue being the shortest and deep red being the longest one. Prisms, or other objects that act as prisms, scatter light into *the visible light spectrum.* It is made up of the colors violet, indigo, blue, green, yellow, orange, and red.

Equipment: Spectroscope or diffraction grating, clear light bulb.

Procedure 1:

1. Look at an incandescent bulb with your spectroscope or diffraction grating.
2. Look at a fluorescent light.

Questions:

1. How are the two light sources different?

_____

_____

2. How are they alike?

_____

_____

Procedure 2:

1. Use the diffraction grating (or spectroscope, or glass slide with slit) to look at the filament of a clear bulb, from about 3 in. away.

Question:

3. Describe the colors you see.

_____

_____

Information: Unlike sunlight, which has all the colors of the spectrum, fluorescent and incandescent light have only partial spectra. Fluorescent lights are rich with blues and greens and lack reds. Incandescent lights have mostly reds and lack blues and greens. This causes problems for people photographing with daylight-type films under these lights. With fluorescent lights a person will appear green. With incandescent (tungsten) lights photographs will come out a brown-orange, reminding you of the old amber-tone prints.

Name _____ Date _____ Period _____

9.6 COLOR COMBINATION AND REFLECTION

Purpose: The purpose of this investigation is to combine colors.

Equipment: Three flashlights, blue, red, and green theatrical gels, plastic wrap, adhesive tape.

Procedure 1:

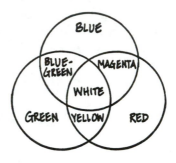

1. Cover the flashlights with red, blue, and green gels. Hold in place with either adhesive tape or plastic wrap.
2. Shine various combinations of these colors on a white screen. Overlap the gels as in the illustration.
3. If a color is too intense in comparison to the others, move the flashlight further away. You should get as follows:

PRIMARY COLORS COMBINATIONS

> red + green → yellow
> red + blue → magenta (purple)
> blue + green → blue-green

blue + red + green → white

Information: When you look at an object, you are looking at the colors it reflects. A green apple appears green, because green reflects while all other colors are absorbed by the apple. Objects absorb all colors other than their own, which they reflect.

Equipment: Red, blue, and green cellophane, filters, or transparent plastic. Red, blue, and green objects.

Procedure 2: Look at red, blue, and green objects through red, blue, and green colored filters or cellophane.

Questions:

1. What color does a green object appear when seen through a green filter? _____.
   When seen through a red filter? _____. When seen through a blue filter? _____.

2. What color does a blue object appear when seen through a green filter? _____.
   When seen through a red filter? _____. When seen through a blue filter? _____.

3. What color does a red object appear when seen through a green filter? _____.
   When seen through a red filter? _____. When seen through a blue filter? _____.

9.7 CHROMATOGRAPHY AND CAPILLARY ACTION

Purpose: The purpose of this investigation is to learn about chromatography and capillary action.

Information: Colors are mixtures of basic dyes. If you wish to paint something, you can select the color and go to a paint store. There they can custom mix your paint. Paint stores have books with recipes for colors. A neutral base paint is mixed with *several* dyes followed by a vigorous shaking to get a uniform color.

Today's project is to examine the components of food colors like the ones you have been using. Food colors, especially the yellow, green, blue, and red, are mixtures of several dyes. The primary dyes are found by using filter paper or paper toweling. Water moves through paper, because paper is porous. This means that is has many fine spaces. This motion of water through fine spaces is *capillary action.* Other examples of capillary action occur in plants, trees, candles, wicks in pens, and kerosene lights. As water moves in a filter, the colors separate into their basic source dyes. The dye separation process is *chromatography.*

Equipment: Paper towels or filters, food coloring, jars, colored pens, ruler, tape, shallow pan, hair dryer (optional)

Procedure:

1. Prepare the filter paper or toweling by cutting it into strips about 4 cm wide. Make them as *long* as the jar with 3 cm to 6 cm to spare.

2. Place one drop of food coloring on the paper, about 1.5 cm from the bottom edge.

3. Place strips inside shallow pan to let food coloring dry, so you will not stain anything else. Let the dye dry or use a hair dryer to speed it up.

4. Prepare all strips, using food coloring, ink from a pen, etc.

5. Label each strip on top with the name of the original color.

6. Place 1 cm of water in jar.

7. Place the prepared strip inside the jar. Tape carefully the top of the strip to the edge of the jar. The dried color spot must be at least 0.5 to 1 cm above the water line.

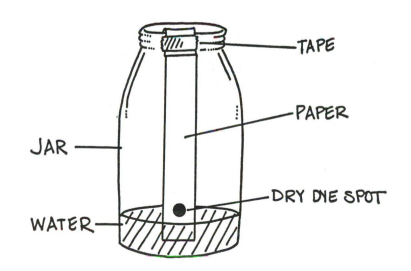

© 1991 by The Center for Applied Research in Education.

213

Sample Data Setup:

Original Color   Separate Dyes

Color Type	Colors Observed

Questions:

1. What dyes make up the red color?

_____

2. What dyes make up the blue color?

_____

3. What dyes make up the green color?

_____

4. What dyes make up the yellow color?

_____

9.8 CAPILLARY ACTION

Purpose: The purpose of this investigation is to investigate the capillarity of different papers.

Equipment: Three or four different brands of paper towels, pie tin, scissors, glass jar, water, food coloring (blue or green).

Procedure 1:

1. Prepare three similar paper strips, one from each type of toweling you have. Cut the strips about 5 cm wide and slightly longer than the height of your jar.
2. Mark two lines on each paper about 8 cm apart, one being 2 cm from the bottom edge.
3. Place each strip in water and measure the time it takes for water to go from the bottom to the top line.

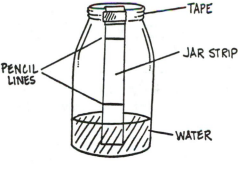

Sample Data Setup:

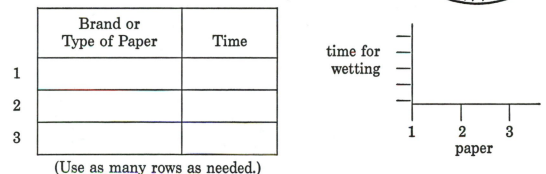

	Brand or Type of Paper	Time
1		
2		
3		

(Use as many rows as needed.)

Procedure 2:

1. Cover the bottom of flat pan with one layer of towel or filter paper.
2. Drop one drop of food coloring in the center of the paper and time for one minute.
3. Measure the spread of the coloring at the widest point and record it.
4. Repeat for two other types of paper or toweling.

Sample Data Setup:

Brand or Type of Paper	Time	
		1
		2
		3

(Use as many rows as needed.)

diameter
of spread
cm

1   2   3
paper

Questions:

1. Rank the papers for absorbency and provide their travel times. Which paper has the largest capillary spaces, which the smallest? How do you know this?

_____

_____

2. Rank the spread of food coloring from the largest to the smallest. Do these data agree with those of water travel? Compare them.

_____

_____

MOTION

TEACHER'S SECTION

In this section the following concepts will be introduced:

1. Speed	5. Pendulums
2. Acceleration	6. Force
3. Potential energy	7. Work
4. Kinetic energy	8. Inertia

10.1 SPEED: MARBLES, ANYONE?

Note: You can choose to do only one or two parts of this investigation. The ramp sizes are only suggested. Use whatever you can get.

Information: This enjoyable activity brings out the best in students. They enjoy your trust and their independence during this highly structured activity. Be careful to use nearly flat surfaces, sometimes impossible. Go through modeling, allowing each group to practice before the real activity. This is an opportunity to discuss variables:

1. Slope of ramps	5. Point where timing is done
2. Launch spot on ramps	6. Size of marbles
3. Levelness of ground	7. Timing method
4. Surface of ground	8. Layout of marking tapes

Borrow stopwatches from the P.E. department, if you do not have them. Most digital watches have a stopwatch function. Assign groups of three or four students per ramp. *Make certain there is a person of at least average ability in each group.* Have the groups decide who will write data, measure time, launch marbles, retrieve marbles, etc. They may trade parts, but keep the same timekeeper. When finished, ask them to return to class (I do this activity outside). Now have them copy the data. Provide them with a calculator to manipulate the data. Ask each student to do his own averages and calculations, to check on the group. I grade the entire group with the same grade. I tell them that the group will get as a grade the lowest score that any team member may earn. This promotes active group support, one of my key aims for this activity.

Average the class results for a grand class average of speeds. Explain to your students that speed is abbreviated with *S* or *s*, while time in seconds is *sec*. [Most] science writers use *s* for seconds (an IEEE and Sl standard), which creates confusion when teaching an introductory unit.

Vocabulary: Speed

10.2 ACCELERATION AND MARBLES

In this activity, you measure the average speed for *specific intervals, 1 meter long, in all cases*. The focus in this investigation is the answers in row C or the *differentials* in average speeds.

Answers:

1. Gravity
2. (Concept: interpolation. You are asking students to provide intermediate values between measurements). Different values will be given, but they should be *decreasing* in value.
3. The differentials are called acceleration; the unit is m/sec².
4. It was decelerating.
5. The rolling friction with the surface.
6. Make ramp steeper, push, make ramp slippery with wax or oil.
7. Roll on surface with greater friction, like shag rug; have it go uphill.
8. Going down the ramp.
9. Between the 0 m and 5 m marks.

Vocabulary: Acceleration

10.3 POTENTIAL AND KINETIC ENERGY: PENDULUMS

In this activity you have two variables:

1. The length of the pendulum string.
2. The masses of the bobs.

In lower and middle grades, students can usually handle only one variable. You may wish to break this fun activity into two, if your group is average or below. Do activity A (in data box) on one day, B on another, and conclusions on the next.

Answers:

1. Potential energy
2. Kinetic energy
3. Gravity
4. Neither; the mass of the pendulum is irrelevant.

5. The shorter one

6. Decrease the string length.

7. Make string longer.

You may want to show a 1 m (39 in.) pendulum—it has a 1 second swing. Use a plumb bob for mass and a sand box.

In 1851, French physicist J.B.L. Foucault hung a pendulum with its bottom sharp point barely touching a sand box. After a few hours, he noticed a change in the direction of the tracks in the sand box. This meant that another force was shifting the motion of the pendulum. This was the first formal evidence for earth's rotation on its axis.

10.4 MOTION, FORCE, AND WORK

Activity 1.6, Objects in Space and Motion, at the beginning of this book, is a prerequisite to this presentation. English units are used at this time, for it is not appropriate to deal with ergs, dynes, newtons, joules, and other related metric units. At an introductory level, one needs to develop concepts before units.

Demonstration: Motion   Ask one student to move across the room. Lift a few objects to illustrate motion. These objects or persons are moving relative to the rest of the classroom or specific observers. Have two students move across the room at same speed. Now they are not moving relative to each other, but they are moving relative to the class. Your reference point determines motion. Ask the class several questions on who is moving, and so forth.

Ask students to provide examples of motion where they do not think or perceive that we are moving. One is the earth moving on its axis. (If you live in Los Angeles, you are now where New York was 3 hours ago.) This varies with your location, be careful with your choice. Other examples: earth's motion around the sun; the solar system moving in the universe; riding in a car with closed eyes; flying in a jet without a reference horizon; riding in a closed elevator.

Conversion Factors: 1 foot-pound = 1.356 joules   1 joule = 0.7376 foot-pound

Answers:

PROCEDURE 2:   Use $W = F \times D$, where W = work (joules), F = force (newton), D = distance (meters). $W = F \times D = .5$ newton $\times 2$ meters = 1 joule

PROCEDURE 3: $W = F \times D = 1$ newton $\times 1$ meters = 1 joule

Vocabulary: Motion, force, work

10.5 INERTIA

Equipment: Table, glass jar with wide mouth, pie plate, hammer or optional broom, styrofoam cup, billiard ball, tennis ball, or hard-boiled egg.

Procedure: You will place a ball in a glass jar by blowing off a plate and ball support between them.

1. Place glass jar about 3 cm to 5 cm from edge of table.
2. Cover jar with solid pie plate. Do not use thin aluminum ones.
3. Turn cup upside-down and center on pie plate.
4. Place ball or egg on top of cup.
5. Place broom vertically by table so that it touches the edge of the pie plate. Step on the broom, bend it back and let go. The broom's handle will hit the pie plate's edge and knock it away together with the cup. The inertia of the ball will hold it in place, and it will drop into the jar.

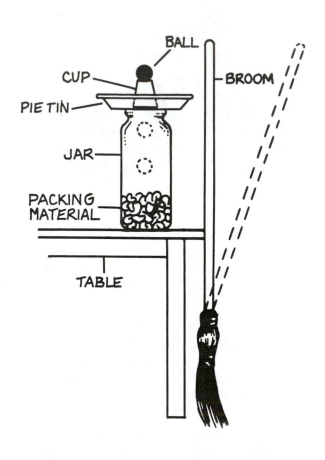

Caution: If your ball is heavy, the glass jar may crack. As a safety step, place some bubble packing, cotton, or foam on the inside of the jar. An alternative is to use a clear plastic jar. Since plastic jars are light and may slide, tape the bottom to the surface of the table.

Notes on Demonstration: When you hit the pie plate, hit only the edge. To have success with the broom method (my favorite), make sure that the broom handle is vertical and touches the table edge. Place the glass jar a short distance away from the edge of the table. The plate should extend past the edge of table. Stand up on the broom, flex it away, and let go. It should hit the pie plate first, then the table's edge, which will stop it. *The heavier the ball, the better.* Practice several times, until you get good at it. You will start a rage.

Other examples of inertia:

1. All stationary objects, until someone moves them.
2. All moving objects: they will slow down on earth, because gravity and friction act on them.

Be careful: if you throw an object in outer space, you will be thrown in the opposite direction. Newton's third law: For every action there is an equal and opposite reaction.

Vocabulary: Inertia

Name _____ Date _____ Period _____

10.1 SPEED: MARBLES, ANYONE?

(Partner activity)

Purpose: The purpose of this investigation is to understand speed.

Information: *Speed* is the distance (length of a path) an object travels in a unit of time. If a car travels 60 miles per hour for one hour, it will have traveled a 60 mile distance in one hour. If a marble rolls 3 meters in 1 second, then its speed is 3 meters per second, written 3m/sec.

Equipment: Five marbles, ramp 6 in. by 12 in. (15 cm by 30 cm), books, stopwatch, meter stick, masking tape

Procedure:

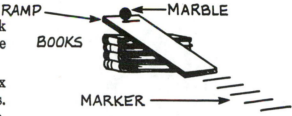

1. Build a ramp with the board placed on a stack of books (25 cm high). Tape the bottom to the ground, to prevent it from moving.

2. Using masking tape, make and install six markers on the floor or across several tables. The first one is at the bottom of the ramp. Mark it *O.* Space the next five at 1 m intervals from each other.

3. Mark the launch point with tape on the ramp.

4. Release the marble from a standing stop. Start timing when marble crosses the zero mark, and stop timing when it reaches the 1 m mark.

5. Repeat five times, record, and average the times.

6. Repeat steps 1 through 5 for the intervals: 0 m to 2 m, 0 m to 3 m, 0 m to 4 m, and 0 m to 5 m.

Sample Data Setup:

PART 1   0 m to 1 m

Try	Travel Distance m.	Total Time sec
1	1 m	
2	1 m	
3	1 m	
4	1 m	
5	1 m	
	Average →	

PART 2   0 m to 2 m

Try	Travel Distance m.	Total Time sec
1	2 m	
2	2 m	
3	2 m	
4	2 m	
5	2 m	
	Average →	

PART 3   0 m to 3 m

Try	Travel Distance m.	Total Time sec
1	3 m	
2	3 m	
3	3 m	
4	3 m	
5	3 m	
Average →		

PART 4   0 m to 4 m

Try	Travel Distance m.	Total Time sec
1	4 m	
2	4 m	
3	4 m	
4	4 m	
5	4 m	
Average →		

PART 4   0 m to 5 m

Try	Travel Distance m.	Total Time sec
1	5 m	
2	5 m	
3	5 m	
4	5 m	
5	5 m	
Average →		

Information: You have just computed the average time for the marble to travel 1 m, 2 m, 3 m, 4 m, and 5 m. To calculate the speed, divide the total distance by the total time. Example: If a marble travels 5 meters in 4 seconds average time, then:

$$\text{average speed} = \frac{\text{total distance}}{\text{total time}} = \frac{5 \text{ m}}{4 \text{ sec}} = 1.25 \text{ m/sec}$$

The average speed of the marble is 1.25m/sec

Questions:

1. What was the average speed of your marble in the 0 m to 1 m interval? If the marble maintains this speed, how far will it go in 10, 25, and 50 seconds?

_____

_____

2. What was the average speed of your marble in the 0 m to 2 m interval? If the marble maintains this speed, how far will it go in 10, 25, and 50 seconds?

_____

_____

3. What was the average speed of your marble in the 0 m to 3 m interval? If the marble maintains this speed, how far will it go in 10, 25, and 50 seconds?

_____

_____

4. What was the average speed of your marble in the 0 m to 4 m interval? If the marble maintains this speed, how far will it go in 10, 25, and 50 seconds?

_____

_____

5. What was the average speed of your marble in the 0 m to 5 m interval? If the marble maintains this speed, how far will it go in 10, 25, and 50 seconds?

_____

_____

6. What happens to the speed of the marble as it travels longer distances? Explain, quoting data.

_____

_____

Name _____ Date _____ Period _____

10.2 ACCELERATION AND MARBLES

Information: A traveling automobile will have to slow down, stop and speed up at times to cope with the traffic. Any change in its speed is *acceleration*. In science this is true for changing to both faster and slower speeds. You use the word *deceleration* in everyday conversation to mean change in speed to a slower rate. In science only positive (faster) and negative (slower) acceleration is used. The unit of acceleration typically is *m/sec/sec* written at times as *m/sec²*. At times you will find *ft./sec/sec* or *ft./sec²*, and so forth. The double time in the unit is your clue that this is a unit of acceleration.

Equipment: Five marbles, ramp 6 in. by 12 in. (15 cm × 30 cm), books, stopwatch, meter stick, masking tape

Procedure 1:   You will be measuring the negative acceleration (slowing down) of marbles. Use a data table like the sample one.

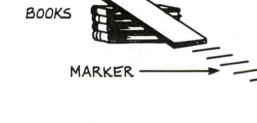

1. Build a ramp with the board placed on a stack of books (25 cm high). Tape the bottom to the ground, to prevent its movement.
2. Using masking tape, make and install six markers on the floor or across several tables. The first one is at the bottom of the ramp. Mark it *0*. The next five markers are evenly spaced at 1 m intervals from each other.
3. Mark the launch point with tape on the ramp.
4. Release the marble from a standing stop. Start timing when the marble crosses the zero mark and stop when it reaches the 1 m mark. Repeat five times, recording each time.
5. Repeat the measurement for the 1 m to 2 m interval.
6. Repeat for 2 m to 3 m, 3 m to 4 m, and 4 m to 5 m intervals.

Sample Data Setup:

	Roll #	Times (seconds)				
		0–1 m	1–2 m	2–3 m	3–4 m	4–5 m
	1					
	2					
	3					
	4					
	5					
A	(average) times					
B	Speed at each mark m/sec → (calculate each)	D	E	F	G	H
C	Difference from previous speed Acceleration m/sec/sec	XXXXXX XXXXXX XXXXXX	D–E	E–F	F–G	G–H

Procedure 2: Do as follows:

1. Average all measured values. Record them in row *A*.
2. Calculate the average speeds in all intervals. Record in row *B*. Example: Distance 1 m, time 2.44 seconds.
   S(speed) = D(distance)/T(time) = 1 m/2.44 sec = .409 m/sec
3. Calculate the differences in average speeds and record them in a row. Notice that each box shows you which values to use.
4. Graph the speed values, plotting the distances on the *X*-axis and the speeds on the *Y*-axis.
5. Graph the acceleration values. Use the *X*-axis for distances and the *Y*-axis for acceleration (changes in speeds).

Questions:

1. What force caused the marble to move down the ramp? Explain.

_____

2. From the graph, what was the average speed of the marble at 1.5 m? 2.5 m? 3.5 m? 4.5 m? 5.5 m?

_____

_____

3. What was the difference in speed from 1 m to 2 m? From 2 m to 3 m? From 3 m to 4 m? From 4 m to 5 m? What are these differences in speed called? What is the correct unit for them?

_____

_____

4. Explain the motion of the marbles from 0 m to 5 m. Was it accelerating, decelerating, or traveling at an even speed? Quote data for support.

_____

_____

5. What made the marble change its speed from 0 m to 5 m?

_____

6. How could you speed up the marble?

_____

_____

7. How could you make the marble roll more slowly?

_____

_____

8. Where was the marble when it accelerated positively?

_____

9. Where was the marble when it accelerated negatively?

_____

Name _____ Date _____ Period _____

10.3 POTENTIAL AND KINETIC ENERGY: PENDULUMS

(Partner activity)

Purpose: The purpose of this investigation is to develop an understanding of potential and kinetic energy, using pendulums.

Information: Energy at rest, ready to be released, is *potential energy.* Released energy in motion is *kinetic energy.* When you lift a ball, the ball has potential energy due to its height from the floor. When you let go, this potential energy becomes kinetic energy (motion). Think of a grandfather clock. A lifted clock pendulum has potential energy. When released, the pendulum moves (kinetic energy). Friction from its bearing and air in the cabinet slow it down. The pendulum would stop were it not for the clock's spring, which provides energy to overcome the friction. The clock spring has potential energy and it is slowly changed into the kinetic energy: the motion of the pendulum. Electricity is potential energy until you throw the switch: then it moves (kinetic energy), making appliances work.

Equipment: String, ruler, pencil, tape, two different lead sinkers, watch

Procedure:

1. Tape pencil to edge of table. It acts as the support for the swinging pendulum.

2. Make two different pendulum strings. The distance from the sinker hook to the pencil is 14 cm with one loop and 28 cm with the second one.

3. Using the 14 cm string, hold the pendulum at the height of the pencil and let go.

4. Count how many times the pendulum swings back and forth (count as one) in exactly 1 minute. Record data.

5. Repeat three times and average.

6. Replace the sinker with the larger one and repeat steps 3 through 5.

7. Replace the string with the 28 cm string and repeat steps 3 through 6.

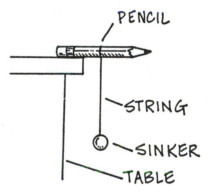

Record all data.

Sample Data Setup:

Pendulum		# of swings in 1 minute			
length of string	sinker	try 1	try 2	try 3	average
A 14 cm	1 oz				
A 14 cm	2 oz				
B 28 cm	1 oz				
B 28 cm	2 oz				

Questions:

1. What energy is in the pendulum before the release? _____
2. What energy is in the pendulum when moving? _____
3. What force causes the pendulum to move when released? _____
4. Which pendulum size causes the pendulum to swing more times?

   _____

5. Which string length causes the pendulum to swing more times?

   _____

6. How could you increase the number of pendulum swings?

   _____

7. How could you decrease the pendulum swings?

   _____

Name _____ Date _____ Period _____

10.4 MOTION, FORCE, AND WORK

(Partner activity)

Purpose: The purpose of this investigation is to clarify the meanings of force, motion, and work.

Information: *Motion* occurs when one material object moves relative to the position of a reference object.

A *force* is simply a push or a pull. When you push a chair or lift a book you use a force. The unit of force is the *newton,* named after Sir Isaac Newton. One newton is the force needed to move a 1 kg mass with the acceleration of 1 meter per second every second.

Scientists define *work* as a force which moves through a distance. Work (joules) = force (newtons) $\times$ distance (meters), or $W = F \times D$. If you push or pull an object with a 2 newton force for 3 meters, then 2 newtons $\times$ 3 meters = 6 newton-meters of work. One newton-meter is called a joule (J). The joule is a unit of work in the metric system. It is named after the scientist James Joule. The joule is used to measure both work and energy.

Procedure 1:

1. Lift a book.
2. Move a pencil.
3. Get up and move your chair a few centimeters.

Equipment: One dish, water, balance, meter stick, 1 kg. mass

Procedure 2: Place container on a balance and fill it with water until it reaches the mass of 500 g. Mark on the closet, wall, door, etc. a 2 m height. Lift the container to that height. Practice until you can do it in 2 seconds.

Procedure 3:

1. Pick up a 1 kg mass.
2. Move it from one end of the meter stick to the other.
3. Do it (with practice) so it takes you exactly one second.

Questions:

1. How much work did you do in procedure 2?

_____

_____

2. How much work did you do in procedure 3?

_____

_____

10.5 INERTIA

Purpose: The purpose of this activity is to demonstrate inertia.

Information: Sir Isaac Newton first stated the laws of motion in the 1600s. His first law of motion states that an object in motion stays in motion unless some force opposes it. He adds that an object at rest will continue to be at rest, unless some force moves it. This law describes *inertia,* the tendency of objects to resist a change in state.

Equipment: Glass jar, coin, piece of cardboard, pencil

Procedure:

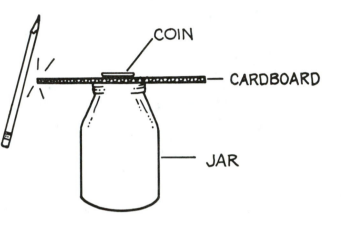

1. Place a piece of cardboard above open glass jar.
2. Place coin on top of cardboard, centered above jar neck.
3. Hit the cardboard with a sharp blow of the pencil.
4. Practice a few times.

Questions:

1. What is inertia? Explain

_____

_____

_____

_____

2. Can you think of other examples of inertia? List them.

_____

_____

_____

_____

GRAVITY

TEACHER'S SECTION

In this section the following concepts will be touched upon:

1. Gravity
2. Center of gravity

11.1 GRAVITY

Demonstration: Gravity and Direction

Equipment: Globe, paper scissors, paper

Procedure:

1. Cut two 2 in. paper figurines.
2. Tape one standing figurine on the north pole, the other on the south pole. Notice that both figures, if they were people, report that down is toward the center of the earth. Neither is aware of being upside down. Neither falls into outer space. Gravity makes this possible.

Answers:

1. The bag stretched the band and moved lower than at the start.
2. Gravity pulled the bag down, because it was heavier.

11.2 FALLING BODIES

The key to these simple activities is to practice so both bodies release at the same time. In Procedure 1, you may substitute a piece of wood (2 in. by 4 in.) for the book. Hit it with a hammer. Make sure that both marbles leave at the same time. (Practice.)

Answers:

1. The marbles fall almost at the same time.
2. Marble falls, while foil floats around.

3. Marble and foil ball fall to the ground almost at the same time.

4. The earth's gravity force pulls on all material objects with the same force. *The force is equal to the mass of the object.* A larger mass will have a larger weight. A smaller mass will have a smaller weight. In vacuum, there are no air molecules and there is no falling resistance from the air, so all bodies fall at the same rate. The mass of an object is the same in either air or vacuum.

Information: A free falling body accelerates (falls faster) every second. The acceleration of gravity is 32.2 ft./sec^2. Due to the drag with air, all falling bodies reach a terminal (maximum) velocity of 450 m.p.h., or 924 km/h. What happens if you have a solid body of water falling? It becomes droplets and mist, due to drag.

11.3 GRAVITY AND CENTER OF GRAVITY

Key concepts to stress:

1. All matter has mass, a property usually described in grams or kilograms.

2. Weight is a property of matter, when affected by the force of gravity.

3. The center of gravity can be at times outside the body or object under discussion. This can lead to instability and unbalance.

On earth hair hangs down, rain falls down, electric lines sag down, cups fall down and break, books fall down. How will they behave in space where there is no gravity? Have students think of many interesting examples. Mention a few NASA problems: to provide meals, showers, toilets, etc., for astronauts.

Answers:

2. On the Moon, one-sixth, on Mars, one-third. On the sun it would be 26 times as much.

3. The c.g. of boys is usually near their chests, while for the girls it is usually by their hips. He or she who bends lower has the lower center of gravity.

4. The c.g.'s are: center of ball, under the wings, between the handlebars and seat, under the chest.

5. Spin the eggs. If they wobble, they are raw. If they spin normally, they are cooked. As a raw egg spins, its yolk moves around, changing its c.g. and causing a wobble.

Demonstration: Center of Gravity   Take a tall narrow box (a wooden fruit crate is ideal).

1. Draw the diagonals on the longest narrow side, their center representing the c.g.

2. Place a cup hook in it and hang a string with a rock, brick, or any mass. The mass should not touch the ground (A).

3. If you tilt the box, while standing on its smallest side, it will return to its normal position without toppling (B). This is true as long as the string falls inside the line of the base below the c.g.

4. The box topples when the *string goes outside the base.* Point out this principle of stability to your students. This is the reason why some people fall off ladders. They lean too far back and their c.g. falls outside the ladder base (the support surface to the ladder).

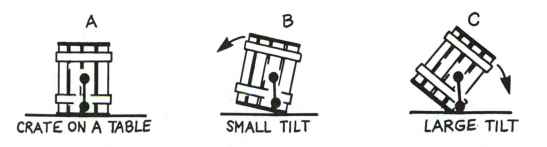

A — CRATE ON A TABLE   B — SMALL TILT   C — LARGE TILT

Many toys and tricks make use of either a lower center of gravity or a broad base to create interesting effects. You can repeat the potato activity with a half-dollar coin and two forks. Place the coin between the forks and balance the coin on the edge of a glass.

Vocabulary: Gravity, center of gravity

11.1 GRAVITY

Information: Gravity is a force that pulls all material objects toward the center of the earth. As one moves away from the center of the earth, the force gradually decreases. Sir Isaac Newton was the first one to describe this force. Weight, a force, is the combination of mass and gravity. If a person travels into outer space, he or she has only mass, but no weight.

Equipment: A small bag, rubber bands, paper clip, books, ruler, marbles

Procedure:

1. Place ruler so that one end is sticking out over the edge of a table. Load several books on the table end to hold it.

2. Cut a long paper strip 1 or 2 inches wide and hang it below the ruler.

3. Place a rubber band over the outside end of the ruler.

4. Tie a small bag with a paper clip and hang it on the rubber band.

5. Place a couple of marbles in the bag. Mark on the paper strip the position of the top of the bag.

6. Add a few more marbles. Mark the new position of the bag.

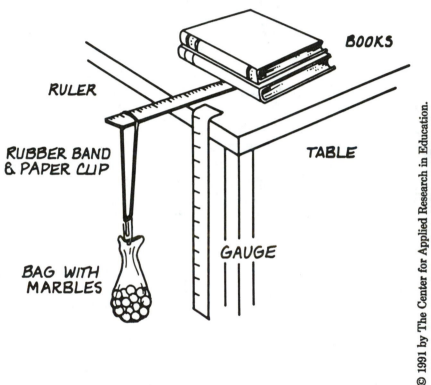

Questions:

1. What happened to the bag when you added marbles?

_____

_____

2. Why did the bag move? What force was acting on it?

_____

_____

Name _____ Date _____ Period _____

11.2 FALLING BODIES

Information: Gravity has the same force or pull on all objects. A one gram mass has a gravitational pull of one gram. A one-kilogram mass has a gravitational pull of one kilogram, and so forth. If two bodies fall at the same time, they arrive on the ground at the same time. Galileo dropped several objects from the leaning tower of Pisa to prove this point. At that time there were no watches. He used his pulse to time the event.

The resistance (drag) of the air does affect the free fall of objects. The greater the surface area, the greater the drag, slowing down the fall. In a vacuum, a feather will fall at the same speed as a lead ball.

Equipment: book, marbles of different size

Procedure 1:

1. Place book on a table, 3 cm from the edge and parallel to it.
2. Place two marbles of different sizes near the book.
3. Push the book evenly and see how the marbles fall to the ground.

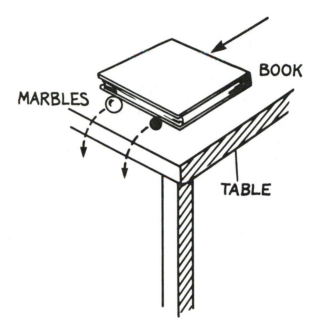

Procedure 2:

1. Drop from your hand a flat piece of aluminum foil and a marble.
2. Squeeze the foil into a small ball.
3. Drop the foil ball and the marble at the same time.

Questions:

1. What happens when you launch two different marbles, as in Procedure 1?

_____

_____

_____

_____

2. What happens when you drop a marble and the foil?

_____

_____

3. What happens when you drop the marble and the foil ball?

_____

_____

4. State the properties of gravity.

_____

_____

Name _____ Date _____ Period _____

11.3 GRAVITY AND CENTER OF GRAVITY

Purpose: The purpose of this investigation is to discuss and to understand the term CENTER OF GRAVITY.

Information: You are a bundle of many different elements in the form of many different molecules. These molecules constitute your mass. On Earth you have weight. This is the attraction or pull of the planet for your mass. Weight, then, is a measure of the pull of gravity. In outer space you keep your mass, but you are weightless! If you climb a tall mountain and measure your weight, you discover that it is less than at sea level. Gravity decreases as you move away from the center of the Earth. Gravity is the amount of attraction or pull one mass has for another mass. The Earth, having a large mass, has a large attraction for you. Sir Isaac Newton defined gravity as the force of attraction between two masses, dependent on the inverse of the distance squared between them. As the distance increases, the force of gravity decreases. At twice the distance, gravity is $\frac{1}{4}$, at 3 times the distance it is $\frac{1}{9}$, at 4 times $\frac{1}{16}$, at 5 times $\frac{1}{25}$, etc. The gravitational pull on the Moon is one sixth, on Mars one third and on the sun twenty-six times that of earth.

EXAMPLES

Imagine the gravitational attraction between two planets of equal size.

d = distance (any very large unit)
f = force of gravity
$M1$ = mass of planet 1
$M2$ = mass of planet 2

$M1 \underset{f\,=\,1/9}{\overset{d\,=\,3}{\rule{2cm}{0.4pt}}} M2 \qquad M1 \underset{f\,=\,1/16}{\overset{d\,=\,4}{\rule{3cm}{0.4pt}}} M2 \qquad M1 \underset{f\,=\,1/100}{\overset{d\,=\,10}{\rule{5cm}{0.4pt}}} M2$

Questions:

1. List several objects and tell how they are affected if gravity is not available, as in outer space.

   _____

   _____

   _____

2. What would be your weight on the Moon? On Mars? On the Sun?

   _____

   _____

Information: If all the mass of an object can be thought as being balanced around one point, this is the *center of gravity* (abbreviated *c.g.*) *C.g.* is the place where you place your hand to lift and balance objects.

Equipment: Ruler, pencil, pen, book, broom, meter ruler

Procedure 1:

1. Balance a pencil or a pen on your fingertip.
2. Balance a pencil or pen on the sharp edge of a ruler. The point where you balance it is the *c.g.*
3. Balance a ruler on its long edge.
4. Balance the ruler on its short edge.
5. Balance a broom or ruler on its end.

Procedure 2:   Find out who has the lower center of gravity; boys or girls.

1. Stand up straight with your back against a wall or door and bend down as far as you can without falling. Measure the height of your fingertips from the floor, just before losing your balance. Your seat and heels must always touch the wall.
2. Repeat several times and average.

Sample Data Setup:

	cm from ground						average
boys							
girls							

Questions:

3. Who can bend lower? Who has the lower center of gravity?

_____

_____

4. Where is the c.g. of a round ball, a jet, a bicycle, a cat?

_____

_____

5. If someone presents you with two eggs, can you find out if they are raw or cooked? Try it at home.

Information: The stability of objects depends on the relationship of their heights and the size of their bases. To make an object stable, you lower its c.g. by reducing its height or broadening its base. Car manufacturers use this principle to stabilize cars to prevent them from rolling over. Pontiac (a division of General Motors Corporation), widened the distance between the wheels of its cars in the 1960s.

Equipment: Two similar forks, one pencil, paper clip, one toothpick, jar or glass, small onion, potato, orange, or clay ball

Procedure 3:

1. Take two forks and stick them into small potato, onion, orange, or clay ball from the opposite sides. Make a letter *V* and bring forks as close together as the neck of the jar or glass will allow.
2. Stick potato with toothpick between the forks.
3. Place toothpick on edge of jar or glass. Balance the assembly.
4. Carefully give it an up and down bobbing motion.

TOP VIEW

Procedure 4:

1. Remove toothpick and stick a pencil through the potato, so its tip is barely sticking out. See illustration.
2. Balance the assembly on the cup edge. With some patience you could balance it on your desk's edge. Give it a try. Give it a gentle nudge and it will bob.

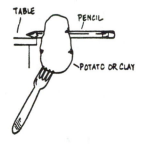

Procedure 5:

1. Open a paper clip and shape it like the letter *U*. Make the three sides about even in length. The sides need to be flat.
2. Push one end of the opened paper clip into the pencil eraser.
3. Place the other end on the edge of a flat object, such as a book cover extended beyond the desk edge.

12

AIR PRESSURE

TEACHER'S SECTION

In this section the following are examined:

1. The weight of air
2. Air pressure
3. Cartesian diver
4. Dams—pressure formula
5. Siphons
6. Bernoulli's principle
7. Newton's third Law

DEMONSTRATIONS: INTRODUCING STUDENTS TO AIR PRESSURE

Equipment: Four red bricks, felt pen, ten textbooks, five sheets of paper, empty metal gallon can with lid (duplicating fluid cans are fine), hot plate or Bunsen burner, water, soap, pot holder, barometer (optional), empty clear plastic or glass soft-drink bottle (2-4 liters), two nails, piece of wire, glass, food coloring, wood board 1 in × 10 in × 3 ft, clear drinking straws, florist clay (or hot glue)

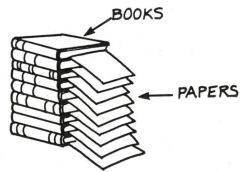

Procedure 1:

1. Stack about ten textbooks, with a sheet of paper between each book. Ask students which papers will be the hardest to pull out.
2. Have several students pull out a sheet of paper from the bottom and from near the top.

Compare air mass and its pressure to a stack of books pressing down on papers. The higher the stack, the greater the weight on the paper. The higher the column of air, the greater the pressure is on the bottom.

Procedure 2:

1. Ask for a volunteer with a smooth, white sole on his/her shoe.
2. Draw a 1-inch square on the sole.

3. Have the student place his/her foot on a chair for everyone to see the square inch.

4. Line up all students and have them pass to each other four red bricks, weighing about 15 lb. Tell them that 14.7 lb./in^2 is the atmospheric pressure that the ocean of air above us presses on every square inch of our bodies.

Information: Relate the increase of pressure as you go swimming in deeper water. The air and water pressures are combined.

As one goes to higher altitudes, the air is not as thick as at sea level. It has fewer molecules per cubic foot and therefore less oxygen. (This is because of less mass above it.) People must breath more air to get the same amount of oxygen. Jets have oxygen masks in case of decompression. Water boils at a *lower* temperature on mountains, due to the decreased air pressure. Some foods, like corn meal, can boil for hours at high altitudes and remain raw. They need a higher boiling temperature than is available to break down their fibers. Pressure cookers can cook foods faster because they increase the boiling temperature of water by increasing the pressure above that of air. Cars have pressure caps on radiators to increase the boiling point of the coolant

Procedure 3: **CAUTION:** Due to increased air pressure inside the can, you can have an explosion, unless you remove the lid.

1. Wash out with detergent an empty one-gallon metal can.
2. Leave a tablespoon of water in it. **LEAVE LID OFF!**
3. Heat can until water vapor starts coming out.
4. Using a potholder, carefully and quickly remove the can from the hot plate.
5. Using the pot holder, quickly replace the cap on the can.
6. Let students see how the can gradually collapses (implodes), after much noise.
7. You can speed up the process by spraying small amounts of cold water on the can.

Notes: You can hang the can on a string with a paper clip hook. Students can see it better. You use water in the can to generate steam, because steam has half the mass as the same amount of dry air. Thus, you have half the number of gas molecules in the can and a much lowered pressure.

Information: Barometers (air pressure meters) show a drop before storms. The air gets moist and has fewer molecules per cubic foot. It weighs less and produces less pressure. Before storms, birds fly lower. Their efforts at wing flapping do not provide them with the usual lift. The air is thinner (far fewer molecules of air per cubic foot). A liter bottle filled with dry air weighs twice as much as when filled with moist air.

Procedure 4: *Assembly of stand.* Cut the board in two pieces: one is 1 ft., the other is 2 ft. Join them at 90° to each other. The 1 ft. piece is horizontal, the 2 ft. piece is vertical. Place a few finishing nails through both.

Bottle assembly.

1. Punch a small hole in the lid of the bottle.
2. Push the soda straw through the lid (¼ in) and seal around it with florist clay or other sealant. (Hot glue works well.)
3. Place lid on bottle.
4. Invert bottle and mount on stand with wire or hot melt glue. The drinking straw *must* be about 1-2 in. from the bottom piece of wood. This allows the insertion of a jar.
5. Place jar under straw. Add water and food color. Make sure that the level of water is several inches above bottom end of the straw.
6. Hold your hands around the bottle. They will warm the air inside the bottle and bubbles will escape through the water. After a while, water will rise in the soda straw. Mark the temperature by reading another thermometer.

The air molecules inside the bottle will come closer and this reduces the inside pressure. Air pressing on the colored water will force the liquid up the straw. This is one of the most sensitive thermometers available. Calibrate it against a known standard, after taping a card behind it.

Procedure 4 (Alternative):   If you have laboratory supplies, use an inverted flask, with rubber stopper and a *long* slender glass rod (3-4 ft.). Place lower end of rod in a beaker with colored water. Support the flask with a stand. Heat flask with a Bunsen burner. You can speed the cooling of the air inside the flask by placing a wet towel on top of the flask or spraying it with a mist of water. Make sure that you are using Pyrex glassware and that students are not close to you. Once in a rare while the flask cracks.

12.1 THE WEIGHT OF AIR

Answers:

1. The meter stick becomes unbalanced when the balloon is inflated with air. Air has mass and weight.
2. The balance is not maintained. The heated air expands and fewer molecules are in the bag. The other bag appears to become heavier.

Note: The balloon with air weighs more. Warm air has fewer molecules, because the molecules are farther apart. Hot air is lighter than cold air and rises. You may want to repeat the previous activity in *reverse* order. Balance a balloon full of air, then let the air go. It works well.

12.2 AIR PRESSURE

Information: Barometers are built with mercury so they need not be 32 ft. high. Mercury rises to 760 mm at sea level. The approximate height of a column of water equivalent to the air pressure is 32 ft. Evangelista Torricelli, (1608–1647) an Italian physicist who was a pupil of Galileo, was important in air pressure studies and thermometer and barometer construction. He used to tour nearby towns and amuse the townsfolk with his contraptions. At times he built barometers 32 ft. high and used church steeples for support.

Answers:

1. The pressure of air holds it there. The bottle can be as tall as 32 ft.
2. The pressure of air is greater than water pressure down.
3. 32 ft.

DEMONSTRATION: EGG IN BOTTLE, BANANA PEELER

Equipment: Hard-boiled egg, vinegar, banana, paper, matches, bottle with neck slightly smaller than egg (flask, juice bottle, or similar)

Procedure:

1. Place a hard boiled egg in vinegar for 10 minutes (to make its skin more elastic), then rinse it with water.
2. Light a piece of paper and while it is burning place it inside the bottle.
3. Quickly seal the mouth of the bottle with the hard boiled egg. The egg enters the bottle with a pop.
4. Repeat and use a partially peeled banana in place of the egg. The bottle will peel the rest of the banana. (This one is messier.)

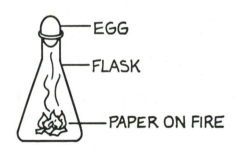

Note: The flame consumed the oxygen, reducing the volume and pressure of air inside the bottle. The air on the outside has more pressure and pushes the egg inside.

To remove the egg from the bottle, turn its neck down, so the egg rests in the neck. Place your mouth over the bottle, form a tight seal and blow hard. KEEP BOTTLE VERTICAL! As you lift the bottle from your mouth the egg comes out.

Blowing air into the bottle creates higher pressure above the egg. The pressure pushes the egg out. Practice these demonstrations, and assign them as homework.

12.3 CARTESIAN DIVER

Note: Ordinary glass eye droppers (as divers) are easiest to make and adjust.

The pressure on the lid transmits to the water which in turn compresses the air inside the diver until both reach the same pressure. Compressed air is heavier than plain air (it has more molecules). Some *water* enters the vial replacing the space used by the air, which is now compressed. This makes the diver heavier and it sinks. The diver is denser than water (due to its mass) and sinks. Observe the rise of the water level inside the vial as the diver sinks. *Pressure* and *volume* of gas vary inversely; as one increases the other decreases. Relate the diver to human swimming and compare human buoyancy when inhaling/exhaling air.

12.4 PRESSURE FORMULA AND DAMS

Answers:

1. Near the bottom, for it has the highest pressure. Think of a stack of books. On the bottom of the stack is the heaviest weight. In the middle it is less. Students will provide distances from data sheet as evidence.

2. Neither, for both have the same height column of water.

3. To compute pressure notice that a 2 cm column of water places a pressure of 2 g over the same 1 cm^2, and so on. A 25 cm high column of water has a force of 25 g over 1 cm^2.

4. $500 \times 64 = 32,000$ lb./ft.^{2} A cubic foot of *sweet* water weighs 62.4 lb., and 1 ft.^{3} contains 7.2 gallons.

DEMONSTRATIONS: SIPHONS

Equipment: Jar, dishpan, one small rubber or plastic hose, water, small chain 4 to 5 ft., food coloring, cup, flask, glass tubing or soda straw, florist clay

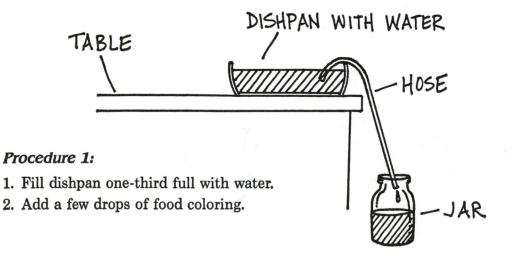

Procedure 1:

1. Fill dishpan one-third full with water.
2. Add a few drops of food coloring.

3. Place pan on table, the jar on the floor.

4. Fill hose completely with water, by placing it in dishpan, under the colored water.

5. Hold one end of the hose in the pan and move the other end pinched closed, *down* into the jar.

6. Open the end in the jar and water will siphon down.

Information: In a siphon there are two parts to the hose filled with liquid: the short one and the long one. The long one is heavier and falls down, due to gravity. It creates vacuum and air pressure forces water up the short end.

Procedure 2:   A simple model which explains a siphon is made with 2 ft. to 3 ft. of chain in a cup. As a small portion of chain overhangs the cup, the chain begins to fall down toward the floor. The mass of the falling chain pulls on the remaining chain in the cup. The chain links hold on to each other just as the electrical charges in water drops do (cohesion).

Procedure 3:

1. Take a flask and cover it with a one-hole stopper.

2. Insert in the stopper a piece of glass tubing (wet stopper). Make certain that the glass tubing goes very close to the bottom of the bottle. Allow about 6 in. outside. You can use drinking straws instead of glass rods, but you must completely seal the hole around it with florist clay.

3. Fill bottle with water, about three-quarters of the way.

4. Blow air into the bottle. Notice the air bubbles rising to the top of the water. Watch the water squirt out.

12.5 BERNOULLI'S PRINCIPLE

The egg in the glass activity works only if you remove the aerator from the faucet. Water goes in and must come out. The speed of water around the egg is high; below it, it is slow. The pressure is low around the sides and high on the bottom. This provides the lift. This is similar to the lift of a large ball by a stream of air. A vacuum cleaner with a narrow air stream adapter (a crevice tool) will do the trick. This is an application of Bernoulli's principle. You can demonstrate this activity and ask students to repeat it at home.

Answer:

1. The bridge bent downwards. The higher air speed below created a lower pressure region.

2. The higher air speed above creates a slight lift.

3. The increased speed of air between the apples creates a lower pressure. The outside air will push the apples together.

12.6 BERNOULLI'S PRINCIPLE AND LIFT

Answers:

1. The front and sides of card are high pressure zones. Behind the card is low pressure. The candle is only an indicator of air currents.
2. The air flows smoothly along the edges of the teardrop or can. It makes low pressure in front and high pressure behind, extinguishing the candle.

12.7 ACTION AND REACTION—NEWTON'S THIRD LAW

The balloon has an erratic (unbalanced) course—it has neither air foils nor a movable exhaust to guide it. Model rockets use their fins for stability. Real rockets control their flights with wings, controlled burn rate engines with gimbaled exhaust nozzles, and small lateral control engines, known as attitude vernier rockets.

When the balloon is full of air, it has potential energy. The air pressure inside it is greater than outside. The air molecules press with *equal* force against all inside surfaces of the balloon. When you let the balloon go, the air escapes through the small opening with quite a force, kinetic energy. An *equal* and *opposite* force to that of the escaping air, pushes the balloon *forward*. This demonstrates Newton's third law of motion: for every action, there is an equal and opposite reaction.

All rockets use this principle. Rockets develop massive amounts of gases by burning their fuel. By limiting the size of the escape hole (exhaust), the velocity of gases is increased, increasing the engine's speed. This is the use of Bernoulli's principle. As the gases escape with great force, an equally great force pushes the rocket forward in the opposite direction. This is Newton's third law.

DEMONSTRATION: BALLOON ROCKET

Equipment: Fine string or thread, drinking straw, balloon, masking tape

Procedure 1:

1. Cut two 2 in. pieces of drinking straw.
2. Place soda straw pieces on a long string and keep them together near one end.
3. Stretch string between the opposite sides of your classroom. Keep it level and taut.

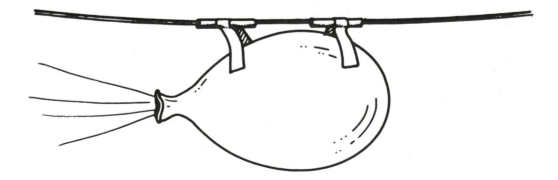

4. Inflate balloon as much as possible. Keep the exhaust closed.

5. Position the balloon ⅛ in. below the string, with its front pointed at the other side of the classroom.

6. Attach balloon with masking tape to the two pieces of soda straw. Students help.

7. Let the assembly go. The balloon will travel across the classroom.

CAUTION: Be careful with the string. It is easy for students not to see it and bump into it.

DEMONSTRATION: LAZY SUSAN

If you have a ball bearing steel turntable (lazy Susan), attach it between two 12 in. square pieces of wood. Occasionally lubricate the bearings (WD 40 T.M. will do).

Let a student stand on the lazy Susan and swing his arms clockwise. Immediately, his body will swing counterclockwise—action and reaction. This is a fun demonstration.

ADDITIONAL ACTIVITY: MODEL ROCKETRY

Build and demonstrate model rockets. They use real solid fuel engines. Estes Industries (Penrose, CO81240) supplies rockets and teaching suggestions. I have taught model rocketry for twenty years. It introduces space technology, and you can use it as the springboard for many science activities. My students are very enthusiastic about it.

12.1 THE WEIGHT OF AIR

(Partner activity)

Purpose: The purpose of this investigation is to learn about atmospheric pressure.

Information: Planet Earth is surrounded by a large ocean of air, which extends up about 500 miles. The bulk of air molecules are on the bottom 75 miles of this ocean. Air appears light or to have no weight compared to material objects. However, a column of air (one square inch at the base) from the surface of the ocean to 500 miles up weighs 14.7 lb./in^2 or 1030 g/cm^2. As you go to higher elevations, air pressure is reduced because there are fewer air molecules above you. You can experience the force and existence of air by slightly extending an arm out of automobile while it is going at 50 m.p.h.

Equipment: Meter stick, string, paper clips, balloon, candle, shallow pan, two paper bags or milk cartons

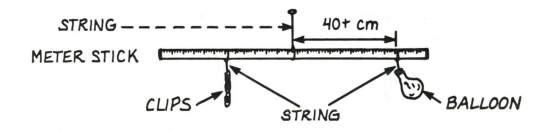

Procedure: Find out if air really has mass.

1. Hang the meter stick from a string.
2. Balance on it two loops of string, one with several paper clips as weights, the other with an empty balloon. Check that the balloon is at least 40 cm from the center of the stick.
3. Carefully remove the balloon, inflate it and hang it back on its string. Be careful not to move the string with the clips. Use tape to hold everything in place.

Question:

1. Does the stick still maintain its balance when the balloon is inflated? Why or why not?

_____

_____

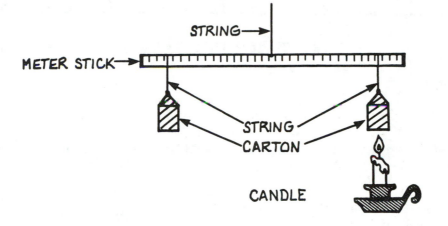

Procedure 2:

1. Balance on the same hanger two paper bags or milk cartons of similar size. Check that the openings face down.
2. Light a candle or bunsen burner under one, making sure that the flame is not touching or burning it. Keep the candle in a shallow pan.

Question:

2. Is the balance maintained? Why or why not?

_____

_____

12.2 AIR PRESSURE

(Partner activity)

Purpose: The purpose of this investigation is to do investigations with air pressure.

Equipment: Bottle or jar, shallow pan, piece of waxed paper, tap water

Procedure 1:

1. Fill the bottle with water and cover it with a piece of waxed paper.
2. Invert and place the bottle into shallow pan half full of water.
3. When the bottle's neck is under water, remove the waxed paper, balance the bottle, and let it stand.

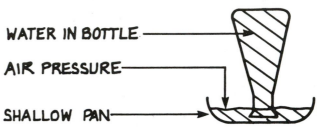

WATER IN BOTTLE

AIR PRESSURE

SHALLOW PAN

Question:

1. Why does the water stay in the bottle, higher than the water level in the pan?

_____

_____

Procedure 2:

1. Fill a jar or glass with water, extra full.
2. Place a piece of aluminum foil over it and lightly press the foil edges against the glass.
3. Carefully invert the glass by supporting the aluminum foil.
4. Remove your fingers and, voilà, water does not fall out!

Questions:

1. Why does the water stay up in the glass?

_____

_____

3. When you drink water or soda with a straw, air pressure pushes the liquids up the straw. You create lower pressure in your mouth, by sucking. What is the longest (straight up) straw that would work?

_____

Information: A tire pump pushes *more* air molecules into a tire than would fit there normally. This action raises the internal pressure of the tire above that of the atmosphere. When you speak of tire or ball pressure, you really speak of its pressure above that of the atmosphere.

Name _____ Date _____ Period _____

12.3 CARTESIAN DIVER

Purpose: The purpose of this investigation is to create a good understanding of air pressure.

Information: Interesting toys and games are made using air pressure. René Descartes (1596–1650) created a brilliant toy which has delighted children of all ages ever since. It is the Cartesian diver.

Equipment: Empty medicine vial or small bottle, quart jar with lid, balloon, 2 or 3 large rubber bands. Alternative: glass eye dropper.

Procedure:

1. Fill small vial with pennies, water, or anything for ballast (weight to make it sink).
2. Cover vial and make a small hole in the lid.
3. Adjust the ballast until the vial, turned upside down, will barely float. Its bottom should barely touch the surface of water.
4. Fill larger jar completely with water and cover it with the lid.
5. Press the lid and the diver will sink.

JAR WITH RUBBER SHEET

— VIAL WITH PENNIES

Alternative Method:

1. Use a large jar without lid.
2. Cover it with rubber sheeting from a balloon.
3. Use sturdy rubber bands to fasten it down.
4. Leave a layer of air about 2 cm to 3 cm or 1 in. above the water.
5. Press on the rubber sheet and the diver sinks. Adjust the ballast inside the diver so the slightest pressure on the rubber sheet sinks it.

Questions:

1. When you press on the jar lid, or on the rubber sheeting, or squeeze the bottle, why does the diver sink? Explain.

_____

_____

_____

2. What happens to the diver when you apply pressure? Explain.

_____

_____

Name _____ Date _____ Period _____

12.4 PRESSURE FORMULA AND DAMS

(Partner activity)

Purpose: The purpose of this investigation is to define pressure.

Information: Pressure has two variables: *height* and the *area* over which it is applied. If you push with your hand against a block of ice, nothing happens. If you push against the same block of ice with a fine pick, the pressure becomes enormous and the block shatters. Using the pick, you concentrate the force of your arm into a small area, increasing it many times.

Cities and towns place their drinking water tanks high in hills, (when possible) to increase water pressure.

When a person weighing 120 lb. dances and has one foot in air, her/his other foot supports her/him. About 3 square inches of her/his sole support him. The floor receives a pressure of 40 pounds per square inch.

When you play a record, a 1 g drop pressure of the diamond needle into the grooves converts into about 40 tons/in² (80,000 lb.). A FPS (Foot-Pound-Second) ton is 2000 pounds. This fact alone accounts for the damage to records when you use them even with the highest quality players. The friction (drag force and its heat) slowly recuts the record grooves. This is not a problem with CDs (compact disks), because there only a beam of laser light scans the grooves. For this reason, they last indefinitely, without quality loss.

Here is the formula for pressure:

$$\text{Pressure} = \frac{\text{Force}}{\text{Area}}$$

A column of air, starting at sea level, puts 1030 g of pressure on the surface of 1 cm² or 14.7 lb./in².

Equipment: Empty half-gallon milk carton, pencil, ruler, dishpan, water, jar, masking tape

Procedure:

1. Make holes in one side of the milk carton. Start at the bottom and go up. Space holes 5 cm apart. Try to make the holes all the same size.

2. Tape all holes with masking tape and fill carton with water, after placing it in a dishpan.

3. Remove the tape from the hole 5 cm from the top.

4. Measure how far water squirts out and record it.

5. Retape the hole.

6. Repeat steps 3 through 5 for all the other holes.

7. Graph values.

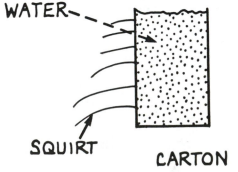

© 1991 by The Center for Applied Research in Education.

Sample Data Setup:

Starting height/level of water in carton_____. Size of carton_____

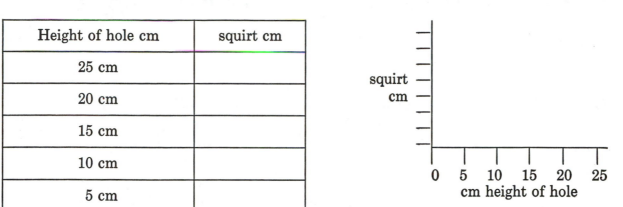

Height of hole cm	squirt cm
25 cm	
20 cm	
15 cm	
10 cm	
5 cm	

Add boxes, if needed.

Questions:

1. If the milk carton represents a water dam, where is the pressure greatest: near the top or the bottom? Why? What is your evidence?

_____

_____

2. You have two different size milk cartons, a half gallon and a quart. Both are filled to the height of 25 cm with water. Both have holes of equal size, spaced at 5 cm from each other. Which carton will squirt the water farther? Why?

_____

_____

3. What would be the water pressure in g/cm² if a carton is filled with 25 cm of water?

_____

_____

4. If a submarine submerges to 500 ft. of depth at sea, what is its hull pressure? (One cubic foot of sea water weighs 64 lb.)

_____

_____

12.5 BERNOULLI'S PRINCIPLE

Purpose: The purpose of this investigation is to learn Bernoulli's principle as applied to flight.

Equipment: paper, two books, jar, water, scissors, string, two apples

Procedure 1:

1. Hold the corners of the narrow side of a sheet of paper, with both hands. Place it just below your lower lip. The paper will fold and bend down.
2. Blow forcefully ahead. See diagram.

Procedure 2:

1. Make a bridge using a piece of writing paper.
2. Place it over two books a few inches apart.
3. Blow below the paper. For stability hold one end of the paper with your finger.
4. Blow above the paper.

Questions:

1. What happened when you blew under the paper? Explain.

   _____

   _____

2. What happened when you blew over the paper? Explain

   _____

   _____

Procedure 3:

1. Hang two apples by a string, so that they are slightly apart.
2. Blow a stream of air between them with a straw.

Question:

3. What happened when you blew the air between the apples? Explain.

_____

_____

Information: Jakob Bernoulli (1654–1705), a Swiss scientist and mathematician, was born in the Netherlands. Bernoulli's principle states that when the speed of a fluid *increases,* the *pressure* on the edge of the stream *decreases.* A *fluid* is a liquid, a gas, or anything that flows. When a large truck passes your car on the highway, it feels as if it pulls you toward it. The passing truck speeds up the air stream around itself, while the edges become a low pressure area. Your car sways toward the low pressure zone.

Earlier you blew air rapidly, creating a low pressure and lifting a sheet of paper. Fluids move from high pressure to lower pressure. Visualize an air pump filling a ball, or water flowing out of a faucet.

Procedure 4:   Water lifts an egg!

1. Place a fresh egg in a large glass of water. The egg will slowly sink to the bottom. Help it down so it does not crack.

2. Place the glass under a water faucet and gradually turn on the water. Try to bring the top edge of the egg as close as possible to the faucet. When the water stream reaches a certain speed, it will lift the egg up against the force of the stream.

12.6 BERNOULLI'S PRINCIPLE AND LIFT

(Partner activity)

Purpose: The purpose of this investigation is to continue examining Bernoulli's principle.

Equipment: Two transparent drinking straws, small jar, water, ruler

Procedure 1:

1. Place one straw in jar with water and use the other one to blow with. Hold straws at a right angle to each other.
2. Flatten the ends of both straws by squeezing them with your fingers.
3. Blow vigorously and keep straws at 90° to each other.
4. Measure the rise of water in the straw. Repeat several times with partner and average the data.

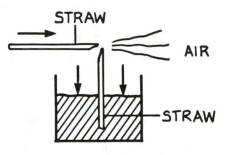

Sample Data Setup:

person	rise of water in straw (cm)								average
	try 1	2	3	4	5	6	7	8	
you									
partner									
							grand average →		

Information: If you blow hard and are careful, water rises in the straw. This illustrates the principle that makes a sprayer work. The lowered pressure at the top of the straw in the jar allows normal air pressure to push water up.

Equipment: Small piece of light board 5 cm to 8 cm square (2 in. to 3 in.), small piece of light board 10 cm by 20 cm (4 in. by 5 in.) (file folders work well as the light board), candle, shallow pan

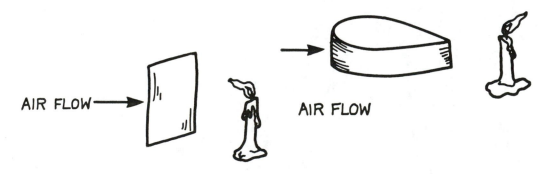

Procedure 2:

1. Cut the light cardboard to the sizes suggested in *equipment*.
2. Light candle inside shallow pan.
3. Place the small cardboard in front of flame and blow forcefully toward cardboard.
4. Take the larger cardboard, fold it in a teardrop shape, or use empty soda can.
5. Point the narrow end at the candle. Blow against teardrop or can and observe.

Questions:

1. What happened when you blew air toward the cardboard? Explain it.

_____

_____

2. What happened when you blew air toward the teardrop or soda can? Explain it.

_____

_____

Information: Notice what happens when a stream of air moves toward the wing of a jet or plane. The lower end of the wing is flat and air moves on at the normal speed. The upper portion of the stream must speed up to travel a longer distance, creating a lower pressure above the wing. Below the wing there is the normal pressure. The difference in pressures creates an upward force called *lift*. Lift is the force that supports the wing. Birds flap their wings to create high and low pressure regions around them even without even knowing of Bernoulli's principle. Flying craft, from simple kites to planes, jets, and space shuttles use this principle.

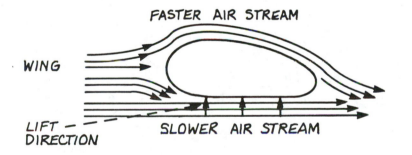

12.7 ACTION AND REACTION: NEWTON'S THIRD LAW

Purpose: The purpose of this activity is to demonstrate Newton's third law of motion.

Information: Sir Isaac Newton (1692–1727) was a giant in science. Newton made many observations about forces, and summarized these in his three laws of motion. He also defined the universal law of gravitation.

First Law: Bodies at rest stay at rest. Bodies in motion stay in motion unless an outside unbalanced force acts on them. *Inertia,* the name for this behavior, is the tendency of all material objects to oppose a change in motion.

Second Law: When an unbalanced force acts on a body, it produces acceleration. This is expressed mathematically as:

$F = m \times a$, where: F = force, m = mass, a = acceleration

Third Law: *For every action, there is an equal and opposite reaction.* If you push against a table, the table pushes back with an equal force. If you push against a closed door, the door pushes back at you with an equal force. If you bulldoze the same door, you break it and go through it the hard way. Your force was greater than that of the door.

Equipment: Balloon

BALLOON

EVEN PRESSURE

OPPOSITE FORCE

ESCAPING AIR (FORCE)

Procedure 1:

1. Inflate balloon and do not let air escape.
2. Let the balloon go while the air in the balloon escapes. Observe and record. Repeat several times.

Questions:

1. Is there a pattern in the flight path of the balloon? Describe it.

_____

_____

2. Why is the balloon moving?

_____

_____